THE EVANS BOOK

LIGHTERS, COMPACTS, PERFUMERS, AND HANDBAGS

Larry Clayton

Schiffer Publishing Ltd

4880 Lower Valley Road, Atglen, PA 19310 USA

DEDICATION

The Evans Book is dedicated to my longtime friend, Kenny Eifert. Kenny has amassed a collection of over six thousand cigarette lighters and is one of the most dedicated lighter collectors in the lighter world. I love you, Kenny.

This book is also dedicated to my beautiful daughter, Melissa Lewis, her husband, James, and my two precious grandchildren, Jessica Ann and Jenna Mae Lewis.

Petite model pocket lighter with matching petite compact. Both in a ruby finish with a fleur-de-lis pattern. $75-$150 set.

Library of Congress Catalog Card Number: 98-86379

Designed by "Sue"
Typeset in *Copperplate Gothic Bd BT/Zurich BT*

ISBN: 0-7643-0641-3
Printed in China
1 2 3 4

Published by Schiffer Publishing Ltd.
4880 Lower Valley Road
Atglen, PA 19310
Phone: (610) 593-1777; Fax: (610) 593-2002
E-mail: Schifferbk@aol.com
Please write for a free catalog.
This book may be purchased from the publisher.
Please include $3.95 for shipping.

In Europe, Schiffer books are distributed by
Bushwood Books
6 Marksbury Avenue
Kew Gardens
Surrey TW9 4JF England
Phone: 44 (0)181 392-8585; Fax: 44 (0)181 392-9876
E-mail: Bushwd@aol.com

Please try your bookstore first.

We are interested in hearing from authors with book ideas on related subjects.

Contents

ACKNOWLEDGMENTS

The Evans book would not have been possible without the extensive research and photography done by Leo and Dot Turini. A special thank you goes to Judith Sanders (Chairman of the Board, OTLS) for her many suggestions and for the support that she generated among OTLS members. Without Leo, Dot and Judith there would not be an Evans book. Thank you, all. I would also like to acknowledge and thank the following people and organizations who have contributed photographs, lighters, or made some other important contribution to *The Evans Book*:

John E. Shoffner
P.O. Box 250
Fife Lake, MI 49633
(616) 897-3912

Dave Irons
Waterford, CT

Mabel Childs
(former Evans Case employee)
North Attleborough, MA

Frederick Marshall
(former Evans Case employee)
Attleborough, MA

Frank Gibson
Cummaquid, MA

Nancy and Andrew Lloyd
Winston-Salem, NC

Kay Latona
(Copy Editor)
Ft. Lauderdale, FL

On The Lighter Side (OTLS)
International Lighter Collectors
P.O. Box 871
Quitman, TX 75783

George Cunningham
Curator, North Attleborough Falls Museum

Beatrice Montplaisir
(former Evans Case employee)
North Attleborough, MA

Catherine Stolman
Upland, CA

Lynn Baker
Tucson, AZ

The Hilsinger Co.
Plainville, MA

Dave Udstuen
Morris, IL

John Elster
Winston-Salem, NC

Deb Witt/Ron Bash
P.O. Box 888271
Grand Rapids, MI 49588
(616) 248-0571

Gay Moreau
Curator, Attleborough Area Industrial Museum

Kesaharu Imai
World Photo Press
Tokyo, Japan

Warren Bushey, Sr.

Mike Stevens
Upper Darby, PA

Tom Sica
Philadelphia, PA

Alan Larrabure
New York, NY

Christine Sanders
Fremont, NE

Ron Eyerkuss
Ofallon, MO

Pocket Lighter Preservation Guild (PLPG)
380 Brookes Dr. #209A
Hazelwood, MO 63042

C. J. Winters
Grass Valley, CA

Ted Ballard
The National Lighter Museum
Guthrie, OK

Terry and Debbie Post
Millersville, PA

Curt Watrouse
Norwalk, CT

Jesse Watrouse
Norwalk, CT

Nick and Julie DiNezza
Amherst, NY

Iver Johnson
(former Evans Case employee)
Largo, FL

Guy Nishida
Los Angeles, CA

INTRODUCTION

While cigarette lighters have been around since the mid-1800s and some people have been collecting lighters for half a century, the hobby did not become organized until about 1984. There are two major lighter-collecting clubs in the United States: On The Lighter Side (OTLS) and Pocket Lighter Preservation Guild (PLPG). Both clubs have international members. There are also several lighter clubs located in Europe. The members of these clubs are interested not only in collecting lighters, but also in the histories of the companies that produced the lighters. Lighter collectors also want to know about other items produced by lighter manufacturers.

The Evans Case Company, an early leader in the industry, is the last of the major American cigarette lighter manufacturers to be documented for historical purposes. While this book is primarily about cigarette lighters, I would be remiss if I did not document some of the other wonderful products that Evans Case manufactured. For that reason, this book also includes beautiful Evans compacts, perfumers, and handbags.

The captions for all items pictured in this book include retail values expressed as a range from good to mint (factory) condition (i.e., $50-150). Because of its rarity, an item in mint condition may be worth many times the value of the same item in good condition. Items in very good or excellent condition will fall somewhere between the good and mint values. While the condition of an item can easily be established, the value is often a matter of personal opinion. The values listed are what a collector can expect to pay for a particular item in today's market.

Chapter 1
History of the Evans Case Company

The Evans Case Company flourished in North Attleborough, Massachusetts, from 1922 until 1960, a period of thirty-eight years. Evans Case manufactured over ten thousand different items, including pocket cigarette lighters, table and desk lighters, cigarette lighter and cigarette case sets, cigarette case-lighters, compacts, vanities, pocket flasks, silver-plated hollow ware, candlesticks, salt and pepper shakers, costume jewelry, plated dresser sets, sterling-silver dresser sets, mesh bags, purses, handbags, watch bracelets, bridge sets and numerous novelties that do not come under any particular classification. The Evans Case Company was a leader in cigarette-lighter manufacturing. Quality control was very rigid. I have never seen a factory-condition Evans cigarette lighter that was not absolutely perfect in quality.

The Evans Case Company started out as the Evans Novelty Company in about 1918. The exact location of production during the early days is unknown but I believe it to have been in the North Attleborough area. The company began with only a half dozen employees and only about a nine-hundred square foot work area. It was a partnership among Alfred F. Reilly, Samuel Haslan, and Fred Burden. Mr. Reilly and Mr. Haslan had become acquainted while both were employed in a manufacturing plant in North Attleborough.

Reilly was a man of enormous talent who had spent nine years of his youth working in various jewelry-manufacturing plants. He had developed considerable skill in metal polishing and electroplating, becoming head of an electroplating division at the age of seventeen, earning $1.20 per day. He became part metallurgist, part chemist, and part engineer, although I do not believe that he had any formal training in any of the production sciences.

Entering North Attleborough.

The Evans Novelty Company began by making picture frames and neck chains (presumably for dog tags) for the United States Army. They did pretty well, making about $3,000 dollars by the third month of production. Shortly after formation of the company, Reilly heard that the United Drug Company was seeking a manufacturer to make a compact to contain its new face powder. He immediately went to Boston, only to find that another company had been chosen to manufacture the compact. Through some inquiries that morning, Reilly determined whom he needed to talk to in order to secure a contract. One account

The former Evans Case plant in 1996.

says that he talked to the General Manager. Another account says that he talked to the President of United Drug. In any case, Reilly showed the individual some samples and was told to bring more samples the following morning. Reilly and one of his partners spent that night making design drawings and plaster samples. He returned to Boston the next day. The customer was very impressed and Reilly was awarded a contract for $250,000 to manufacture the compacts.

Reilly had an order in hand but did not have a plant or machinery for manufacture of the compacts. He bought some secondhand machinery and looked for a place to set it up. Reilly talked to the owners of the Totten Building at 21 East Street in North Attleborough. The Totten Building had been constructed in the early twentieth century and at that time was home to some dozen or so jewelry manufacturing businesses. He secured space on the first floor and began to install his machinery. About this time, the United Drug Company decided to increase its order. Worried that Reilly might have problems filling the increased order, United Drug sent several senior officials on an unannounced visit to the building on East Street. Mr. Reilly took the officials on a grand tour of the whole building, giving the impression that all of the jewelry manufacturing businesses in the building were a part of the Evans Novelty Company. The officials went away with broad smiles and confidence in the company's ability to deliver the product. A few days later, Reilly returned to Boston to ask for a cash advance on the order. His request was immediately granted and he walked away with a substantial amount of start-up money.

Business was going so well that the company incorporated, becoming the Evans Case Company in 1922. Alfred Reilly became Chairman of the Board of Directors at this time. He eventually bought out his two partners, Haslan and Burden. At some point in time, he installed his three sons as officers in the company. William J. Reilly became President, Alfred F. Reilly, Jr. became Executive Vice-President and Donald Reilly became Vice-President of Sales. I believe that all three sons are now deceased. The business continued to flourish. During 1922, the company had about seventy-five employees. Floor space had increased to thirty-five hundred square feet.

Early on, Alfred Reilly began experimenting with cloisonné enamel finishes for a wide range of products. It was difficult to compound a type of enamel that would bond fast with steel alloy and have the same coefficient of expansion as the metal. Reilly studied, developed, and eventually perfected the enameling process. As a result of working out a successful formula, no Evans cloisonné enamel lighter or compact will chip or break when dropped on the floor.

Frank E. Nolan, who had a background in manufacturing, joined the company as General Manager in 1926. By 1928, the company had begun to manufacture cigarette lighters. By 1932, it had expanded to four hundred employees and sixty thousand square feet of work area. Sales volume reached one and a half million dollars. All male labor worked a thirteen-hour workday. The company's seasonal peak was between August and December of each year, when the work force expanded to seven hundred employees with a weekly payroll of approximately twenty thousand dollars. This was the largest individual payroll in the North Attleborough area at the time. By 1935, employment had reached one thousand during the seasonal peak, the company's all-time-high employment figure.

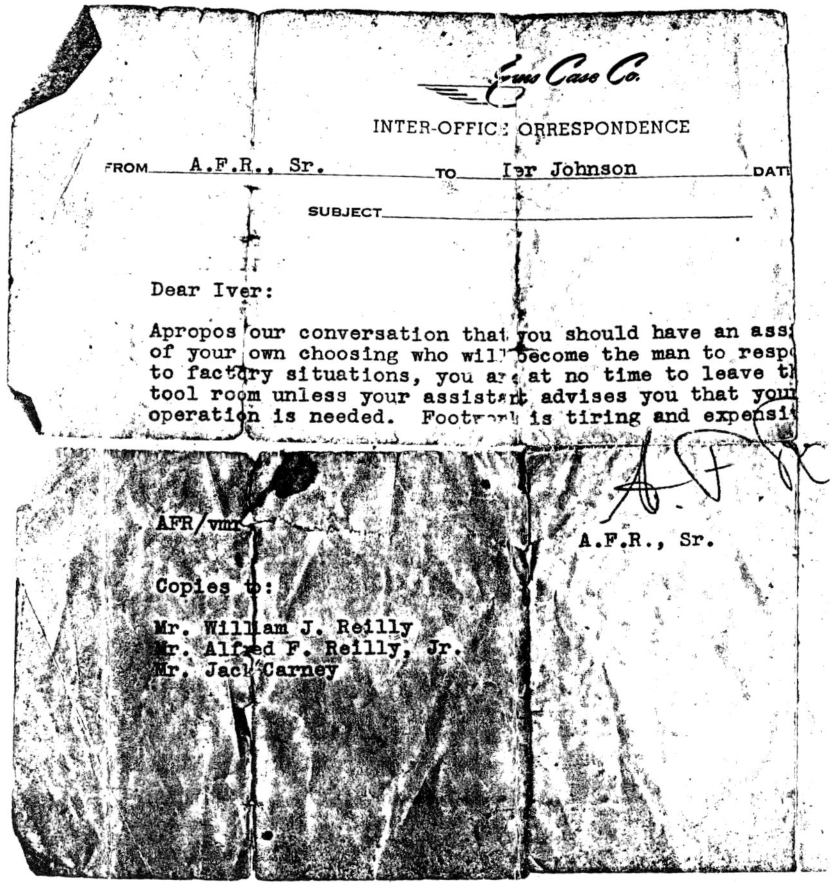

Evans Case Co.

INTER-OFFICE CORRESPONDENCE

FROM A.F.R., Sr. TO Iver Johnson DATE

SUBJECT

Dear Iver:

Apropos our conversation that you should have an ass of your own choosing who will become the man to resp to factory situations, you are at no time to leave th tool room unless your assistant advises you that your operation is needed. Footwork is tiring and expensi

AFR/vmr

A.F.R., Sr.

Copies to:

Mr. William J. Reilly
Mr. Alfred F. Reilly, Jr.
Mr. Jack Carney

Memorandum from Alfred Reilly to Iver Johnson sent sometime during the 1930s. Mr. Johnson was the Tool Room supervisor at Evans Case for many years and currently resides in Largo, Florida.

Alfred Reilly spent considerable time on the road selling the company's products. In between trips he could be found at the plant sitting at his desk, developing some new design or trying to solve some difficult production problem. Mr. Reilly was a master at methods of cost cutting and thoroughly scrutinized every step of a manufacturing process to find ways to save time and money. He made his rounds of the factory on a regular basis. Employees remember that he always had a pleasant smile and a friendly, "Hi, how goes?"

In 1932, a company named Art Metal Works (later known as Ronson) filed suit against a department store and the Evans Case Company for selling cigarette lighters that infringed on their patent for the "one motion" lighter. The Evans Case Company was the manufacturer of these cigarette lighters. Art Metal Works (AMW) won at the trial level, but the decision was appealed by Evans Case and reversed in 1934 by Judge Martin T. Manton of the United States Circuit Court of Appeals. Not until 1939 did the whole story of the reversal come out. A young New York prosecutor had been researching unusual happenings in appeals over which Judge Manton had presided. He presented his findings to a Grand Jury and Judge Manton was indicted for accepting bribes. Soon after the indictment, Manton resigned from office. During the judge's trial, Alfred F. Reilly testified that he had bought the decision from Judge Manton for $25,000 which he had passed to William J. Fallon, Manton's go-between. Reilly paid Fallon an additional $25,000 by putting him on the Evans Case payroll.

After Manton's trial was over, the United States Supreme Court reversed his Evans Case decision and once again Art Metal Works had sole rights to the "one motion" lighter. The only problem was that, due to judge Manton's decision, the patent had been unenforceable from 1932 until the Supreme Court's reversal in 1939. In 1944, the Congress of the United States, in an unprecedented action, granted Art Metal Works, Inc., a seven-year extension on its patent number 19023.

When the Supreme Court reversed in favor of Art Metal Works, Inc., AMW was in a position to completely destroy Evans Case by forcing it into bankruptcy resulting from the latter's inability to sell any more of the lighters that infringed on AMW's patent. Had AMW chosen this path, it is likely that it would have received little in payment from Evans Case for

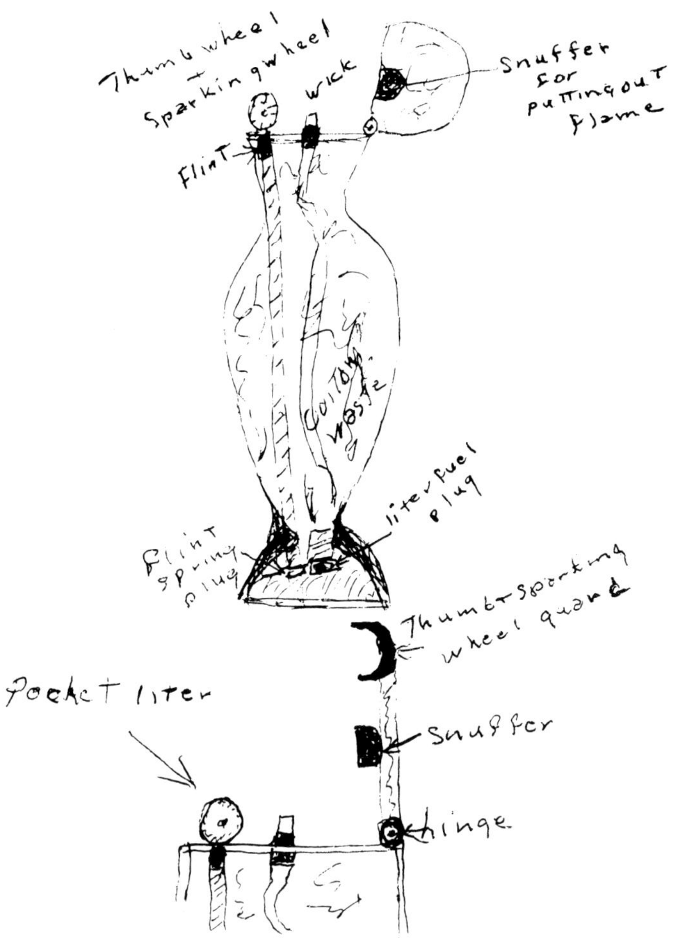

Sketches of early Evans table and pocket lighters drawn in 1990 by Mr. Frederick Marshall. Mr. Marshall worked in the Hub and Die Room at Evans Case from 1934 until 1960. He currently resides in Attleborough, Massachusetts.

the patent infringement. AMW chose, instead, to make the best of the situation for both parties by entering into an agreement with Evans Case. The Evans Case Company agreed to pay AMW a $50,000 licensing fee and 5 percent of gross sales for the use of the "one motion" patent until it expired on June 12, 1952. This explains the appearance of AMW's patent number on Evans lighters beginning in the early 1940s. The expiration of the patent no doubt explains why Evans Case totally redesigned its pocket lighters in the early 1950s.

During the years of World War II, Evans Case curtailed production of compacts and jewelry items. However, cigarette-lighter production continued, to ensure that U.S. servicemen stationed in far-off lands would have a means to light their cigarettes, as dry matches were hard to come by. Handbag production also continued. Handbag frames were made of hand-finished wood because of restrictions on the use of metal products. Employees numbered about four hundred and wages averaged fifty cents per hour.

In early 1943, the company received a War Department contract to manufacture over one million incendiary bombs. The bombs weighed about six pounds each and were bound in five-hundred-pound clusters to be dropped over Japan by B29 bombers during the war. Evans Case also manufactured surgical instruments for the Army and Navy as well as precision parts for aircraft engines, tools for the production of aircraft ammunition, radar components, and aircraft engine identification plates.

In May of 1943, Evans Case hired guards and installed wire fences around the plant to protect classified work it was doing for Sylvania Electric Products. Later, in October of 1945, it was revealed that Evans Case had been involved with manufacturing components for Sylvania's radio proximity fuse. This fuse was installed inside anti-aircraft artillery shells to detonate the shell when it came close to enemy aircraft.

The years between 1946 and 1950 were the greatest era for Evans Case. The company had opened sales offices in New York, Los Angeles, Chicago, Boston, Cleveland, and Dallas. The total work force was about six hundred. Payroll had reached $850,000 annually. In 1946, Alfred Reilly embarked on a major expansion to the plant that would make Evans Case the largest manufacturer in North Attleborough. Approximately 21,600 square feet of work area was added for new polishing, electroplating, and lacquering rooms. The new polishing room was equipped with seventy-five polishing heads, which more than doubled polishing capacity. The new electroplating room was outfitted with enough tank capacity to hold five thousand gallons of plating solution. An acid-proof floor was installed and the electroplating room was set up for mass production. The plant expansion brought the total work area up to approximately 100,000 square feet. The entire structure of the old building had to be raised to the level of the new additions. Because of the high water table of the area, workers soon reached water while piledriving six huge (4 foot diameter) metal caissons to provide new support for the old building. Deep-sea divers in full gear were called in to help with the job. They descended into the caissons to pump out sand and debris and prepare the caissons for the pouring of concrete.

In 1949, Drew Pearson reported in his Merry-Go-Round column that Alfred Reilly was threatening to demand a Congressional investigation of General Douglas McArthur. At that time, General McArthur was responsible for the rebuilding of post-war Japan. Mr. Reilly charged that McArthur was encouraging production of cheap Japanese cigarette lighters to the point that American cigarette-lighter manufacturers were in serious jeopardy. Reilly charged that the Japanese had made copies of Evans, Ronson, and other American lighters, using copper molds and thus saving research, tool-and-die designing, and other costs that American companies had borne.

In 1950, cigarette-lighter production at Evans Case averaged twelve thousand lighters per day. The company advertised widely in all of the leading magazines such as *Vogue*, *Harper's Bazaar*, *Town & Country*, *House Beautiful* and *House & Garden*. Every Evans advertisement of the time carried the slogan, "Evans is Elegance." All of the leading department stores clamored to sell Evans products. Few manufacturers of lighters and style accessories got as much window display and counter space as did Evans Case. Fashion editors and the editors of women's pages in the big dailies gave space, without stint, to pictures and descriptions of beautiful Evans handbags and carryalls. Evans Case products were featured on many radio and television give-away shows.

The company had a Ceramics Department for making its own ceramic bases for table lighters. In the 1950s, it produced many table lighters made of fine bone china. In the late 1950s and early 1960s, ceramic bases with Evans inserts were also manufactured by Royal Doulton and the Balfour Company. Also, some ceramic table-lighter bases were made in France. According to one former employee, table-lighter bases made of art glass were produced by someone other than Evans Case.

There is some evidence that Evans Case owned a perfumer factory located in Brooklyn, New York. Pictured in this book, under the section on perfumers, is a set consisting of a table lighter and matching perfumer, both in clear acrylic with a yellow rose pattern. Both items are labeled on the bottom, "Evans Atomizers, Inc., 62 Stanhope Street, Brooklyn, New York."

The company held a prestigious reputation in the jewelry industry and was known for the mechanical perfection, jewelry finish, and high quality of its products. There were numerous quality inspections for every item manufactured. Orders poured in and it was very difficult to keep up with the demand. The Design Department worked three years ahead of current production. Evans designers continuously observed style trends in the leading cities of America. By the time a new item had been designed and was ready to go into production, it was a foregone conclusion that it would gain wide public acceptance.

During the early 1950s, the Evans Case employees were unionized by the Retail, Wholesale, and Department Store Union, Playthings, Jewelry, and Novelty Workers Division of the United Jewelry Workers, Local 591, CIO. Prior to that, employees had received an annual four percent bonus plus gifts of household appliances and other items. Arrival of the union appears to have signaled the beginning of the end for Evans Case. The contract between Evans Case and the union came up for renewal in the spring of each year. Throughout the 1950s, newspapers reported strike threats by the union, although the employees never did go on strike. After the union came in, Evans Case began a gradual reduction of its work force, possibly as a way to cut costs and maintain its profit margin in the face of the union's demands for higher wages. Many employees were laid off. Evans Case could not maintain previous levels of production utilizing this smaller work force. Evans cigarette lighters manufactured during the mid- to late 1950s have always been much harder to find than lighters made only a few years earlier, reflecting dramatically decreased production.

Sometime around 1955, Alfred Reilly's wife decided that she wanted to run the Handbag Department. As a result of this decision, two of the senior managers in the Handbag Department quit outright. This threw the whole Handbag Department into disarray. Soon after, the Handbag Department was shut down permanently. Sixty employees were laid off.

Butane lighters were made in the Evans Case plant during the mid-to late 1950s. According to one former employee, some workers were deathly afraid of the butane, with good reason. Someone's carelessness could have blown up the whole plant.

Mr. Reilly had moved to West Palm Beach, Florida, in a semi-retired status. Operation of the plant was left to his son William. There apparently was some friction between William and the elder Reilly; whenever William started the development of a new product line, his father sent word back to North Attleborough for the development to cease.

The Evans Case Company was sold to Aaron Krock, a Worcester banker and industrialist, on March 10, 1960. Mr. Reilly came up from Florida and met privately with Mr. Krock. When the meeting was over, the company, including the building, had been sold for about one million dollars. It was reported that a Canadian factory owned by Evans Case was also a part of the sale. What products the Canadian factory made is unknown. There were several reasons for the sale, according to a firsthand report from a surviving family member. Alfred Reilly was in his sixties and ready for retirement. His wife wanted him to remain retired in Florida. He had grown weary of the constant pressure from the union for more pay and benefits. I believe that he decided to sell instead of passing ownership to William because he did not trust William to run the company on his own. William had no idea that his father had been thinking about selling the company and was heartbroken when he heard the news of the sale. Newspaper accounts of the time stated that "no personnel cuts are anticipated" but that "various alterations and changes in operating procedures and sales distribution will be made as soon as possible." It was announced that handbags would be completely redesigned and once again be made available. William J. Reilly would be involved "partly with management, but mostly with sales and designing." Alfred F. Reilly, Jr., would be offered a position in sales, as would Donald Reilly.

Despite the optimistic newspaper reports, Mr. Krock apparently soon changed his mind about continuing operation of the company. He may have decided that he did not want to face union negotiators during contract renewal talks that were scheduled for late April, 1960. Another possibility is that he never intended to keep the company operating but just saw it as a candidate for liquidation that could turn him a profit. In any case, during early April, 1960, there was a newspaper report that the company would shut down for three to four weeks in order to consolidate all operations on the first floor. The report noted that space on the second and third floors would be rented to other businesses. Production was completely shut down by mid-May of 1960 and all production workers were laid off. Production was never resumed. The company was reduced to a skeleton staff of about thirty-five people. Most of the production equipment was sold to the Doran & Johnson Manufacturing Company. Evans Case continued to sell from its retail store and to fill orders from its very large inventory. In June of 1960, Aaron Krock publicly announced that the Evans Case Company was for sale.

In November of 1961, Mr. Krock sold the Evans Case name, product line, and distribution to the Hilsinger Corporation of Plainville, Massachusetts. The Evans building was not a part of the sale. The Hilsinger Corporation is still in business in Plainville today and is a manufacturer of optical lab products and optical accessories for use by ophthalmic professionals. There is also a line of consumer/patient products sold under the Hilco brand name.

William Reilly joined Hilsinger and the company continued to sell the Evans Gift Line items, table lighters, and pocket lighters for several years until the stock received from Evans Case was almost depleted. Hilsinger manufactured butane lighters under the Evans name to supplement the stock received from Evans Case, according to an individual who was employed by both Evans Case and Hilsinger. Evans advertisements of the 1960s, actually placed by Hilsinger, showed the Evans location as Plainville. Evans cigarette lighters could be returned to Plainville for repair. Sometime during the 1960s, Hilsinger contracted with the makers of the "Prince" lighter (a Japanese model) to manufacture an inexpensive butane lighter known as the Eyelighter. This lighter is almost identical in appearance to butane lighters of the time that were marked "Prince." The lighter was marketed by Hilsinger in the United States.

In 1962, the Evans Case building at 21 East Street was sold to the Balfour Realty Trust. Balfour had been in business in North Attleborough for many years. The company had an immediate need for substantial manufacturing space to produce its ceramic, crested knitwear, and specialty items. It is not known how long Balfour remained in business at the East Street location. Today, the former Evans Case building houses several small businesses. The owner of the building is a lumber company located on the premises.

Hilsinger closed its lighter division about 1971. Any and all remaining Evans stock and repair parts were turned over to Authorized Repair Service, 30 West 57th Street, New York, New York 10019. Authorized Repair Service continues to sell and service Evans cigarette lighters to this day.

In 1996, I visited North Attleborough and stood across the street from the former Evans Case building. I took pictures of the building and knew that I was capturing a part of history . I went inside so that I could say to my Evans collector friends, "I have been inside the building where Evans cigarette lighters were made." How many Evans collectors can say the same?

Chapter 2
North Attleborough Falls Museum

George Cunningham, Curator, North Attleborough Falls Museum.

The North Attleborough Falls Museum is located at 100 Commonwealth Avenue, Attleborough Falls, Massachusetts, 02763 in a building that once housed the Attleborough Falls Fire Barn. The Fire Barn was built in 1893 to support local firefighting efforts and closed in 1930. The museum was opened in 1993 by its curator, George Cunningham, and contains over seven thousand articles of local history of the North Attleborough and Plainville areas. It is significant to the research and writing of *The Evans Book* because it displays approximately one hundred items related to the Evans Case company, including Evans cigarette lighters, compacts, jewelry, and handbags. Two of the Evans display items are pictured here. The museum also contains newspaper articles and magazine advertisements related to the Evans Case Company and Evans cigarette lighters.

The second floor of the building houses the North Attleborough and Plainville Chamber of Commerce. The present museum has been visited by over forty-five hundred school children and has had visitors from as far away as England and Spain. Anyone desiring information related to the history of local industries may visit the museum or contact the curator, Mr. Cunningham, at the above address. The museum is open on Saturdays from 9AM until noon.

Attleborough Falls Fire Barn.

Evans Banner model pocket lighter with enameled floral pattern. $15-50.

Bracelet with white enamel finish and dark-colored stones. $20-$40.

Chapter 3

Catalog Pages and Advertisements

EVANS 1929

EVANS 1929

Courtesy of Kesaharu Imai, Lighters, *World Photo Press, Tokyo, Japan, 1997.*

Courtesy of Kesaharu Imai, Lighters, *World Photo Press, Tokyo, Japan, 1997.*

EVANS 1930

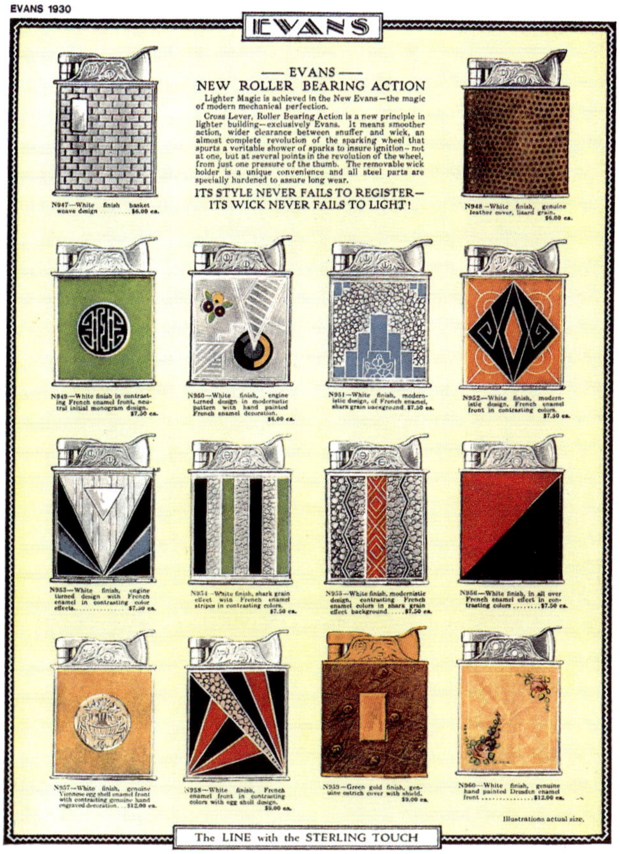

EVANS 1929

EVANS 1929

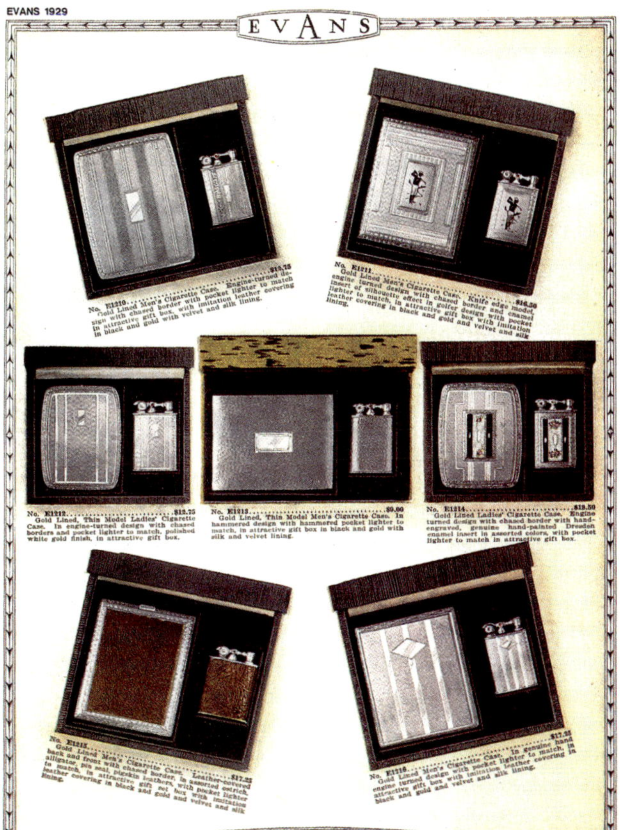

EVANS 1929

Courtesy of Kesaharu Imai, Lighters, *World Photo Press, Tokyo, Japan, 1997.*

EVANS 1929

EVANS 1930

EVANS 1930

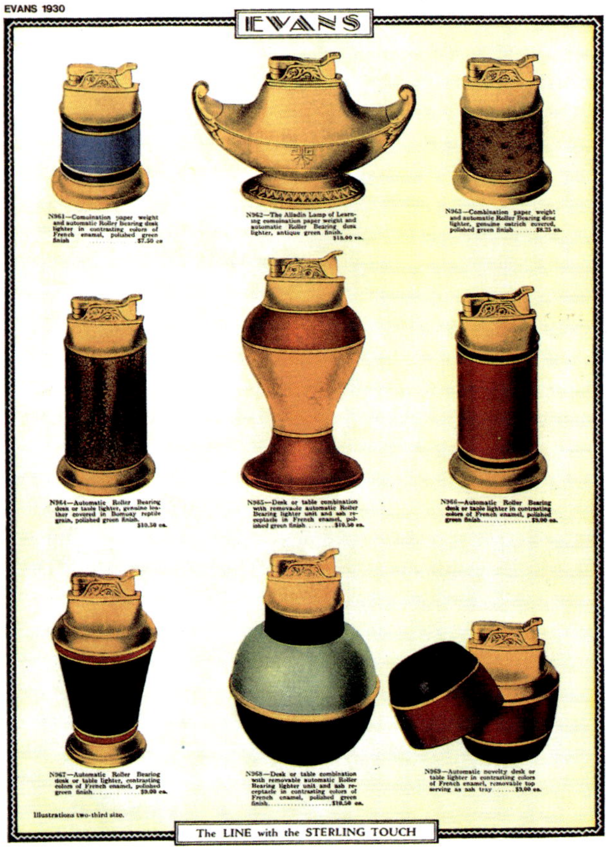

EVANS 1930

EVANS 1930

EVANS

EC300—White finish, hammered and engine turned design. ....$3.00 ea.

EC301—White finish, genuine leather cover, ostrich grain. ....$3.75 ea.

EC302—White finish, shark grain effect with modernistic design in contrasting French enamel colors. $4.50 ea.

EC303—White finish, French enamel front in contrasting colors..$6.00 ea.

EC304—White finish, modernistic design with French enamel front with egg shell effect. $7.50 ea.

EC305—White finish, engine turned design with contrasting French enamel decoration...$6.00 ea.

EC306—White finish, modernistic design, French enamel front with egg shell effect. .....$7.50 ea.

EC307—Green gold finish with sterling silver top front in genuine hand painted Cloisonne enamel. $9.00 ea.

EC308—Green gold finish, sterling silver top front in genuine Cloisonne enamel. ....$9.00 ea.

EC309—White finish, French enamel front with genuine Cloisonne enamel decoration. ....$7.50 ea.

EC310—White finish, Automatic Roller Bearing Desk or Table Lighter, French enamel front with genuine Cloisonne enamel decoration. ....$10.50 ea.

EC311—Big Boy model, Automatic Roller Bearing Pocket Lighter, white finish engine turned design with French enamel stripes, maximum fuel capacity for pocket lighter. ....$6.00 ea.

EC312—White finish, Sport model, Automatic Roller Bearing Pocket Lighter, modernistic design, French enamel front with egg shell effect. ....$9.00 ea.

ILLUSTRATIONS FOUR-FIFTHS ACTUAL SIZE

World's Largest Manufacturers of Style Accessories

Courtesy of Kesaharu Imai, Lighters, *World Photo Press, Tokyo, Japan, 1997.*

INTERNATIONAL SILVER, 1930

EVANS 1930

INTERNATIONAL SILVER, 1930

Courtesy of Kesaharu Imai, Lighters, *World Photo Press, Tokyo, Japan, 1997.*

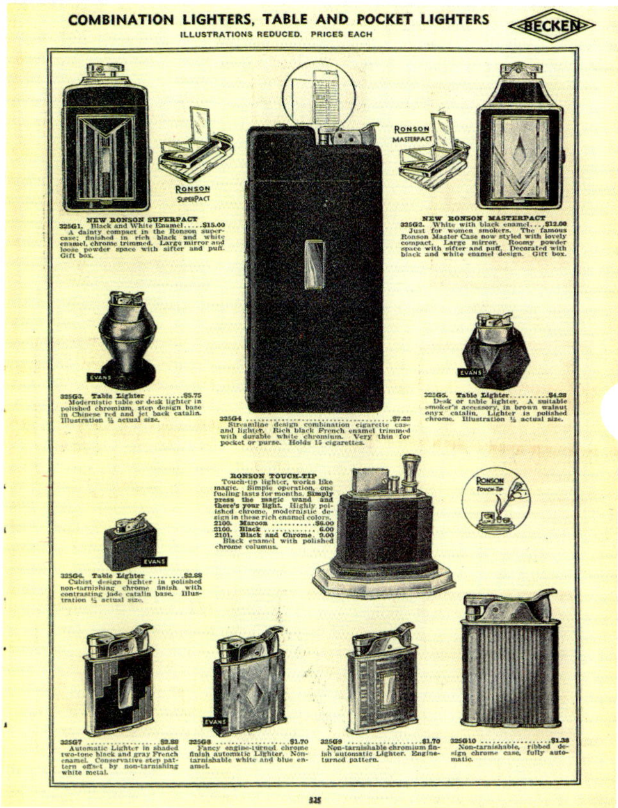

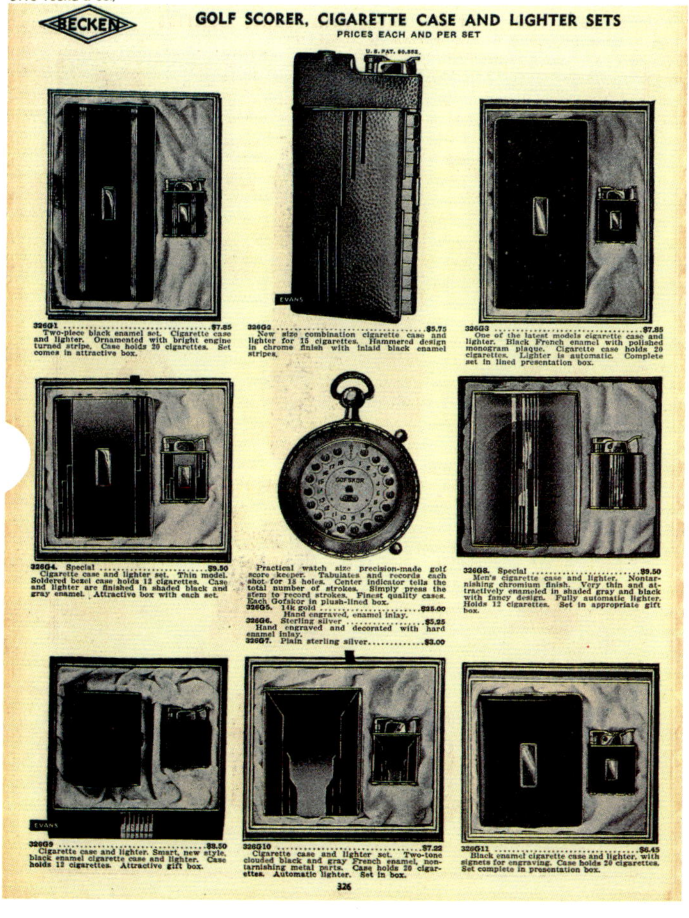

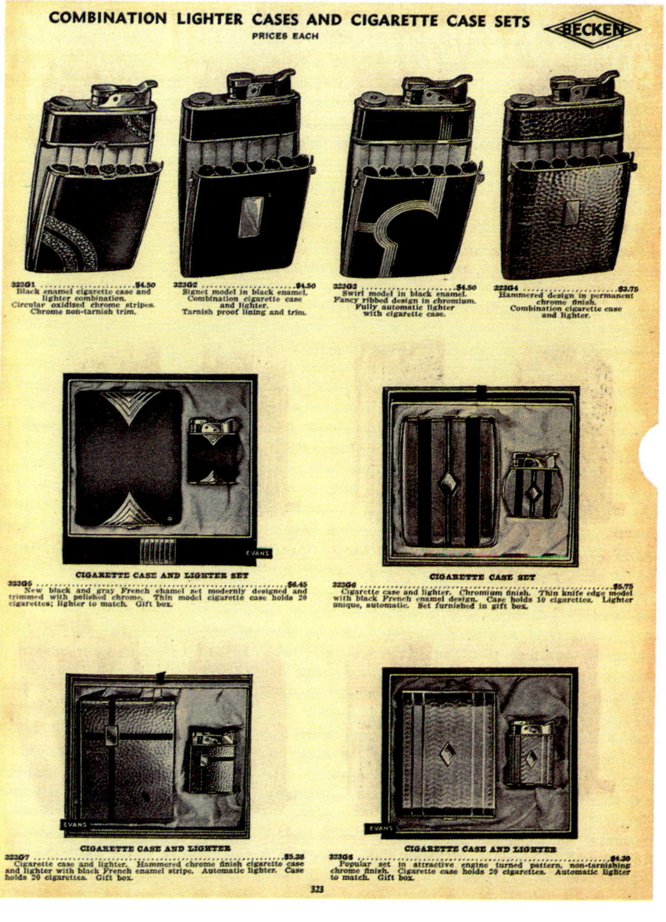

Courtesy of Kesaharu Imai, Lighters, *World Photo Press, Tokyo, Japan, 1997.*

OTTO YOUNG & CO., 1939

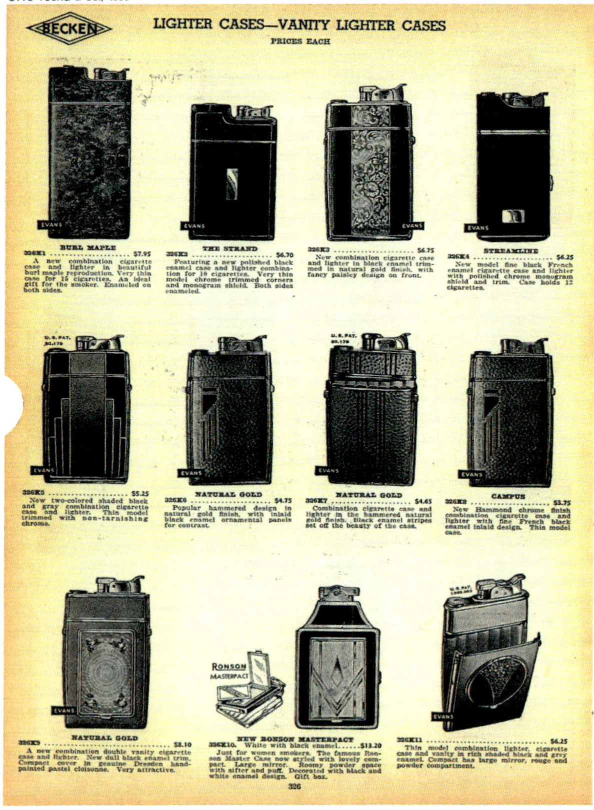

OTTO YOUNG & CO., 1939

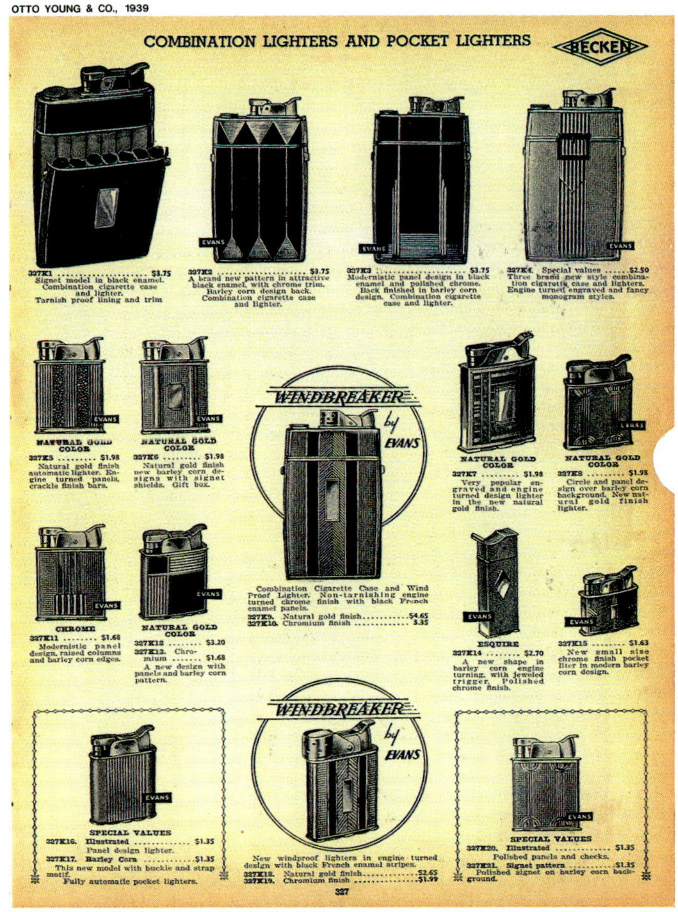

L.&C. MAYERS CO., 1939

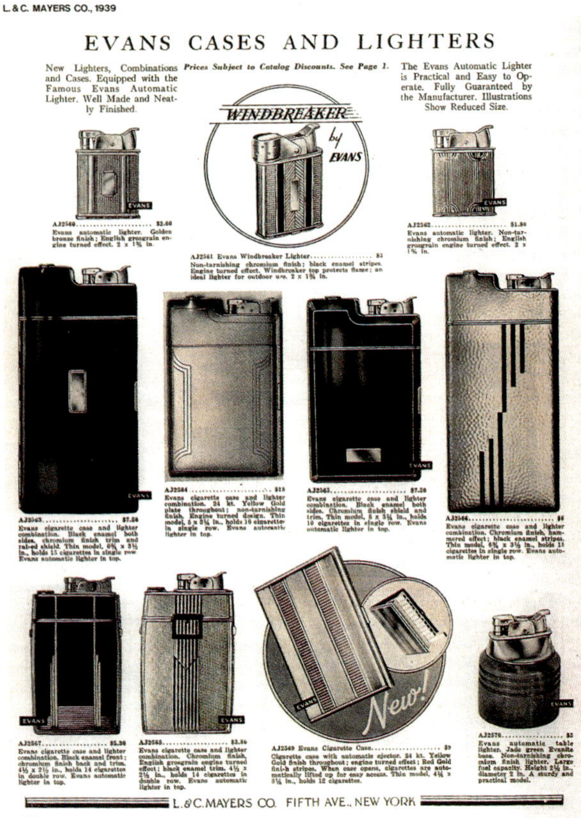

OTTO YOUNG & CO., 1939

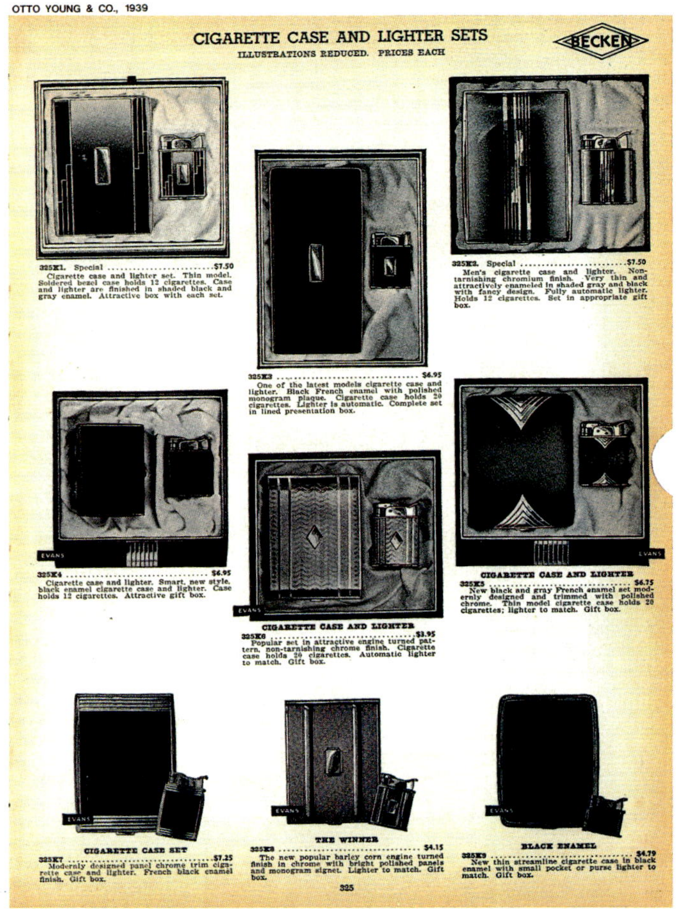

Courtesy of Kesaharu Imai, *Lighters*, World Photo Press, Tokyo, Japan, 1997.

EVANS PASTEL ENAMELS

From the magical world of EVANS fashion fresh accessory combinations. Rainbows of softly shimmering jewel-like pastel tones. Breathtaking effects previously executed only on precious metals now brought to lustrous life in EVANS Pastel Enamels. Once again proof that EVANS IS ELEGANCE.

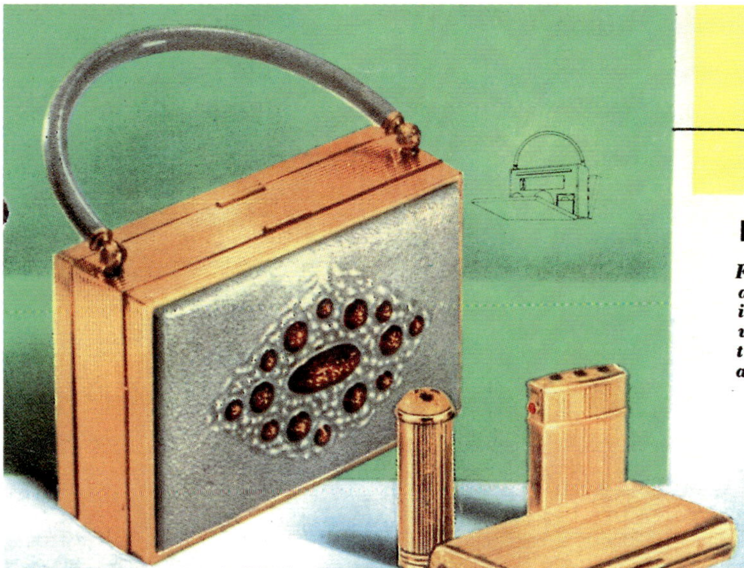

Magnificent evening case in "Jewelers Bronze" and grey Enamel touched with ruby. Her every need served with famous fittings; lighter, lipstick, feather touch compact, comb, mirror, kingsize cigarette case, money purse and quarter coin receptacle.

EC150................Retail **$37.50** Plus Tax

Deep 3 dimensional amethyst combines with tan Evans Pastel Enamel and 14K gold electroplate on this rich accessory set. Exclusive feather touch compact.*

EC151	Lighter, Cigarette Case Lipstick .... Ret.	**$16.00**
EC152	Lighter only .... Ret.	**5.00**
EC153	Cigarette Case only.... Ret.	**7.50**
EC154	Lipstick only .... Ret.	**3.50**

Bright sapphire blue moments captured on sparkli straw Evans Pastel Enamel. Dimensional butterfli lite softly on gay charming purse accessories. Exclusi feather touch compact.*

EC155	Compact, Lipstick, Mirror, Comb . . Ret.	**$15.**
EC156	Compact only	**6.**
EC157	Makeup mirror only	**5.**
EC158	Lipstick only	**3.**
EC160	Lighter, Cigarette Case	**15.**
EC161	Lighter only	**7.**
EC162	Cigarette Case only	**7.**

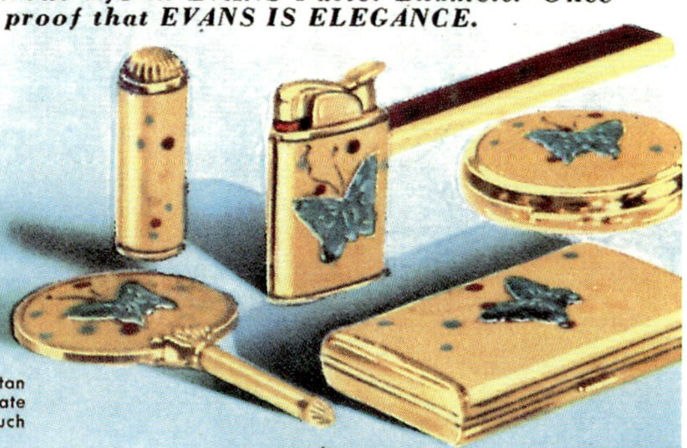

Petal pink and pearl-like Evans Pastel Enamel shades intricate flower brocade pattern accented with encrusted 24K gold. Exclusive feather touch compact.*

EC163	Lighter, Cigarette Case, Compact ....Retail	**$21.00**
EC164	Lighter only	**7.50**
EC165	Compact only	**6.00**
EC166	Cigarette Case only.....	**7.50**

Treasures in Evans Enamel. Ladies razors in tan with gay accents or dainty rose on white. Full-size head takes popular blades.

EC167	Tan ....Retail	**$5.00**
EC168	White ....Retail	**5.00**

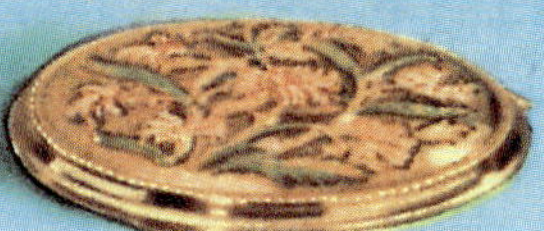

Something really "nice" in traveling mirrors. Generous double mirrors feature both regular and magnifying sides. Wire stand provides rest.

TRAVELING MIRRORS

Ever popular soft subtle Pastel Evans Enamel Ming Pattern or Basketweave Pattern in flawless "Jewelers Bronze."

EC169	Ming Pattern Retail	**$10.00**
EC178	Basketweave Pattern . . Ret.	**$7.50**

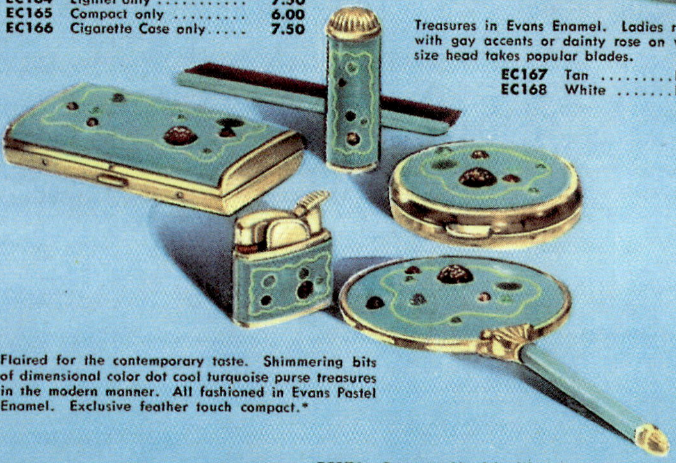

Flaired for the contemporary taste. Shimmering bits of dimensional color dot cool turquoise purse treasures in the modern manner. All fashioned in Evans Pastel Enamel. Exclusive feather touch compact.*

EC170	Compact, Lipstick, Mirror, Comb ....Retail	**$18.00**
EC171	Compact only	**6.00**
EC172	Makeup mirror only	**7.50**
EC173	Lipstick only	**3.50**
EC175	Lighter, Cigarette Case	**15.00**
EC176	Lighter only	**7.50**
EC177	Cigarette Case only	**7.50**

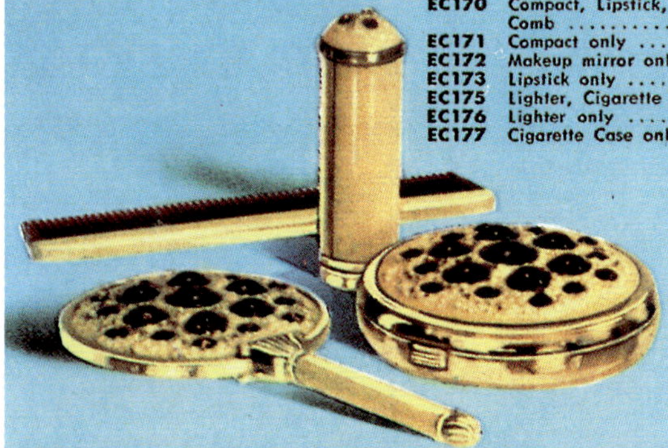

Classic 3 dimensional topaz lends distinction to straw Evans Pastel Enamel on "Famous Fashion Accessories." Exclusive feather touch compact.*

EC179	Compact, Lipstick, Mirror, Comb......................Retail	**$19.50**
EC180	Compact only	**7.50**
EC181	Makeup mirror only	**7.50**
EC182	Lipstick only	**3.50**

An enticing "New Look" for evening, this golden satchel in softly shinin textured finish crowned with daring bits of gay intriguing enamelled quartz Faille interior pockets.

EC184 .... Retail **$25.0**

***A light tap at the center of the Feather Touch compact releases puffs of your personal mixture of loose powder. Ends messy powder spillin**

EVANS FINE CHINA

Masterpieces of unexcelled beauty in color design and function by the leading manufacturers of Fine China smoking sets. Inspired combinations of golden "Jewelers Bronze" metal and fine china in a wide range of hues and tones produce home fashion accessories sure to suit any taste.

The entire body of each Evans Fine China Table Lighter serves as a fuel reservoir for many months supply.

All illustrations one-half actual size.

Graceful flowing lines and bright spatters of white and gold on pleasing eggshell create a **"Thing of Beauty"** in this **"Pitcher"** lighter with complementing king size cigarette box and 5½" ashtray. Flaired thumb-lever provides easy lighting action. Also in black, specify.

EC255 3-piece set ....Retail $20.00
EC256 Lighter only 15.00

Evans all time favorite the **"Quilt"** set. Soft eggshell white accented by bright dots of gold. A beautiful addition to the boudoir. King sized cigarette box and 4½" ashtray. Also in pink, specify.

EC257 3-piece set...Ret. $25.00
EC258 Lighter only 15.00

Fascinating four piece Evans Fine China smoking set in tastefully decorated artful 24K gold spatter work on soft eggshell black. Also in white, specify.

EC259 Lighter, cig. urn, two 3" ashtrays..Retail $12.50

Now a **"New Look"** to the famous **"Fish"** set in Fine China by Evans. Tall gracious lighter and complementing cigarette urn and 5½" ashtray, startling in cloudy grey swirls, and tiny gold fish on jet black.

EC260 3-piece set ....Retail $17.50
EC261 Lighter only 12.50

Reminiscent of Arabian Nights encrusted in 24K gold. Your wish for a conversation piece come true. **"The Magic Lamp"** in Evans Fine China.

EC263........ ....Retail $12.50

EVANS ENAMELS–

JEWELRY OF THE HOME — *Compellingly beautiful are stately "Coffee Table Sets" fashioned in glistening transparent Evans Enamels. Exclusive techniques, unmatched craftsmanship and artistic ability combine to produce enamelling effects previously possible only on precious metals.*

Entire body of each lighter serves as a long lasting fuel reservoir.

All illustrations one-half actual size.

The oriental flavor of the **"Ming Tree"** in deep red 3 dimensional Evans Enamel contrasts richly with gorgeous green. A striking touch in any setting.

EC236 Lighter, cig. urn, 6" ashtray......Retail $21.00
EC237 Lighter only 10.00
EC238 Compote—6½" 15.00

Deep brown 3 dimensional Evans Enamel butterflies flutter enticingly over sun bright pastel straw. A spot of color for any room. 4½" ashtray.

EC242 2-piece set ....Retail $15.00
EC243 Lighter only 10.00

Regal and luxurious in topaz 3 dimensional jewel-like Evans Enamel on pastel straw crowned by 24K gold encrusting. **"Priceless"** in appearance.

EC239 Lighter, cig. urn, 6" ashtray......Retail $28.50
EC240 Lighter only 15.00
EC241 6" Bon Bon Dish 15.00

Reborn again in life-like 3 dimensional Evans Enamel acorns and subtle tan the famous **"Acorn"** set. A perfect complement to all decors. King size cig. box and 9½" dec. tray.

EC244 3-piece set ....Retail $35.00
EC245 Lighter only 10.00

Evans

Unmatched beauty and dependability from Evans . . . the fashion leader in fine home and personal smoking accessories.

Sensational Super Breeze King, America's smartest windlighter. Automatic lighting action provides surefire always. Sparkling polished chrome finish and ribbed design.

EC-100..Retail $3.95

King size cigarette case and famous Evans automatic lighter in engine turned design on gleaming golden metal. Engraving shield perfect for monogramming. Also available in polished chrome finish. Specify. Gift packaged.

EC-106 2 piece set..........Retail $6.00 EC-107 Lighter only..........Retail $3.00

Ladies Lighter in Evans genuine cloisonné enamel. Small and compact, just right for pocket or purse.

EC-101.........................Retail $5.00

The Super Breeze King in handsome black alligator calf and brilliant polished chrome, featuring easy one hand action.

EC-102.........................Retail $5.00

Super Breeze King in rich genuine snake and golden metal. Long lasting fuel supply.

EC-103.........................Retail $6.00

For a thoughtful gift—The Super Breeze King. Smart new model with attractive wood graining. Golden metal shield suitable for engraving.

EC-104.........................Retail $7.50

Evans automatic lighter and pop-up "Serv-Pac" cigarette case in colorful plaid makes this set something special. Gift packaged.

EC-108.......Retail $6.00

"THE EXECUTIVE"—Distinctive Desk Set, perfect for office or den. Has removable automatic lighter and Fine China ash tray. Illustrated in simulated black morocco. Also available in brown. Specify.

EC-109..........................Retail $10.00

Evans genuine cloisonné enamel men's accessory set. Distinctive Warrior design on matching lighter, cuff links and tie bar. Gift packaged.

EC-105.........................Retail $10.00
Plus $ .50 Tax

Evans

EVANS FINE CHINA

SENSATIONAL NEW COMBINATIONS OF F
CHINA AND GLISTENING GOLDEN METAL FC
FAMOUS DECORATOR STYLED SMOKING S
BY EVANS.

Entire body of each lighter is a lc
lasting fuel reservoir.

A conversation piece in any home—this clever Evans automatic table lighter in shape of Castle Chessman. Combined with matching cigarette box and ash tray, an outstanding smoking set. Illustrated in Palomino Beige. Also available in White. Specify.

EC-119 3 piece set....................Retail $25.00
EC-120 Lighter Only ....Retail $15.00

Simple beauty reflecting the grace of your home. A classic three piece smoking set in the Evans tradition of Fine China. Also available in black or white. Specify.

EC-121 3 piece set
Retail $17.50
EC-122 Lighter Only
Retail $10.00

The Hitching Post—novel and handsome—table lighter, ash tray and cigarette box in soft black eggshell finish. Also available in Palomino Beige or White. Specify.

EC-123 3 piece set..........Retail $20.00
EC-124 Lighter only..........Retail $12.50

Hand decorated in the contemporary manner. Three piece smoking set in black with golden Fish design on Evans Fine China.

EC-125 3 piece set....................................Retail $15.00

A brand new idea in smoking sets. Styled to s
the most modern taste and decor. Black wrou
iron contrasts with attractive tri-colored automa
lighter, cigarette urn and ash tray. Also availa
in all white. Specify.

EC-126 3 piece set....................Retail $10.

All sets attractively gift packaged. Illustrations approximately one-half actual size.

Evans
FINE BONE CHINA TABLE
LIGHTERS AND SMOKING SE
Finely fashioned accessories designed
complement any home. Conversat
pieces in the rich tradition of china. T
crowning touch to home decor.
ENTIRE BODY OF EACH LIGHTER IS A FUEL RESERVOIR.
ALL SETS ATTRACTIVELY GIFT PACKAGE
All Illustrations ½ Actual Size
Quilt Design
EC-63 Retail $25.00
Lighter Only
EC-64 Retail $15.00
Above also available in Black or White. Specify.
Antique Auto Stein Set
EC-67 Retail $15.
Lighter Only
EC-68 Retail $10.
Contemporary American
EC-65 Retail $20.00
Lighter Only
EC-66 Retail $12.50
Above also available in grey. Specify.
Shadowbox with Currier and Ives prints.
EC-70 Retail $15.00
Modern Table Setting
EC-69 Retail $15.00

EVANS
Evans brings you Fine Bone China combined with gleaming metal in a new concept of decoration. Produced through an exclusive process, hand decorated by Evans artisans . . . accessories of enduring beauty.
RED SEAL FUEL CONTROL PREVENTS EVAPORATION OF FLUID.
All Illustrations ½ Actual Size
Rosette Swirl Design
EC-71 Retail $25.00
Lighter Only
EC-72 Retail $15.00
Gold-Toned Modern Design
EC-73 Retail $12.50
Above also available in White. Specify.
Early American Motif
EC-74 Retail $25.00
Lighter Only
EC-75 Retail $15.00
Melon Fluted Design
EC-76 Retail $18.00
Lighter Only
EC-77 Retail $15.00
Above also available in Black or White. Specify.
Swedish Contemporary
EC-78 Retail $15.00
Above also available in Black or White. Specify.
ALL SETS ATTRACTIVELY GIFT PACKAGED.

EVANS
. . . superb craftsmanship brings the age old art of fusing enamel and metal to modern day perfection. Preserving unexcelled beauty in a lasting functional form.
RED SEAL FUEL CONTROL PREVENTS EVAPORATION OF FLUID.
ALL SETS ATTRACTIVELY GIFT PACKAGED.
Rosettes on White 3 Piece Set
EC-28 Retail $27.50
Lighter Only
EC-29 Retail $12.50
Above also available in black. Specify.
Trojan Horse
EC-31 Retail $10.00
Dragon
EC-30 Retail $10.00
Ming Tree 2 Piece Set
EC-32 Retail $22.50
Lighter Only
EC-33 Retail $10.00
Petit Point Pattern
EC-34 Retail $17.50
Topaz Colored Enamel Stoned. 3 Piece Set
EC-35 Retail $35.00
Lighter Only
EC-36 Retail $15.00

EVANS WOMEN'S PERSONAL FASHION ACCESSORIES DISTINCTIVE AND GLAMOROUS GIFTS BY AMERICA'S FOREMOST STYLISTS.

STUNNING NEW ITEMS IN GLITTERING RHINESTONES ON DELICATE FILIGREE APPLIQUE AGAINST LUSTROUS GOLDEN BACKGROUND.

EC-127 Fully fitted carryall*..........Retail **$22.50**
EC-128 King Size Cigarette Case..Retail **$12.50**
EC-129 Automatic Lighter..Retail **$10.00**
EC-130 Compact ....Retail **$10.00**
Above Items Plus Tax

Luxurious carryall in stunning textured design on golden metal. Embellished with full cut marquise rhinestones in captured settings. Fully Fitted.*
EC-131..............Retail **$17.50** Plus Tax

Delicate engine turned design graces this popular Evans fully fitted carryall.*
EC-132..............Retail **$12.50** Plus Tax

Charmingly conceived is this splendorous Tu-Toned golden metal brocade pattern.
EC-133 Automatic Lighter ....Retail **$5.00**
EC-134 King Size Cigarette Case........Retail **$5.00**
EC-135 Compact ....Retail **$5.00**
EC-143 Three Piece Set, gift packaged..Retail **$15.00**

Soft and Durable Evans flexible mesh in rich golden metal finish.
EC-136 Eyeglass Case ....Retail **$7.50**
EC-137 Ladies Wallet .... Retail **$8.50** Plus Tax
EC-138 King Size Cigarette Case.............................Retail **$10.00**
Also available in regular size—specify
EC-139 Coin Purse...Retail **$5.00** Plus Tax

Captivating hand decorated golden Crown pattern on black in Evans genuine cloisonné enamel.
EC-140 Automatic Lighter.......................Retail **$6.00**
EC-141 King Size Cigarette Case........Retail **$6.00**
EC-142 Compact ....Retail **$6.00**
EC-144 Three piece set, gift packaged..Retail **$18.00**

*All Evans carryalls contain automatic lipstick, coin purse, powder compartment, large make-up mirror and space for King or Regular size cigarettes.

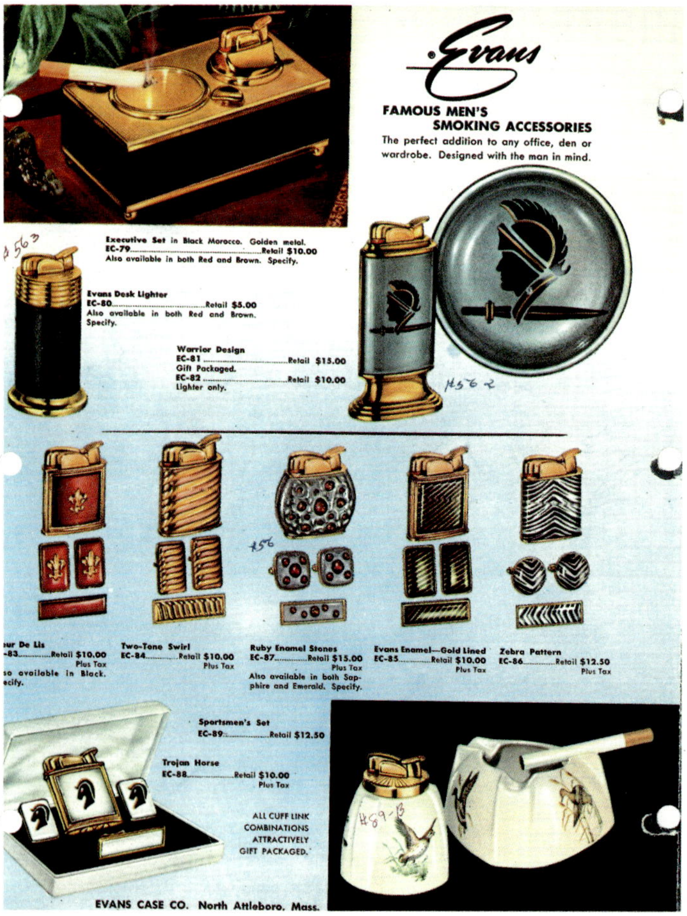
Evans
FAMOUS MEN'S SMOKING ACCESSORIES
The perfect addition to any office, den or wardrobe. Designed with the man in mind.
Executive Set in Black Morocco. Golden metal.
EC-79 Retail $10.00
Also available in both Red and Brown. Specify.
Evans Desk Lighter
EC-80 Retail $5.00
Also available in both Red and Brown. Specify.
Warrior Design
EC-81 Retail $15.00
Gift Packaged.
EC-82 Retail $10.00
Lighter only.
Two-Tone Swirl
EC-84 Retail $10.00 Plus Tax
Ruby Enamel Stones
EC-87 Retail $15.00 Plus Tax
Also available in both Sapphire and Emerald. Specify.
Evans Enamel—Gold Lined
EC-85 Retail $10.00 Plus Tax
Zebra Pattern
EC-86 Retail $12.50 Plus Tax
Sportsmen's Set
EC-89 Retail $12.50
Trojan Horse
EC-88 Retail $10.00 Plus Tax
ALL CUFF LINK COMBINATIONS ATTRACTIVELY GIFT PACKAGED.
EVANS CASE CO. North Attleboro, Mass.

DISTINCTIVE NEW ENAMELED COMPACTS AND COMBINATIONS
Strikingly different designs handcrafted in the exquisite Evans manner.
EVANS
Rose Design on White lighter and compact.
EC-49 Retail $8.00
Also available in Chartreuse. Specify.
Sunburst Design on golden metal lighter, lipstick and compact.
EC-47 Retail $10.00
Poodle lighter, compact, lipstick and cuff links.
EC-50 Retail $17.50 Plus Tax
Rosettes on Black lighter and lipstick-compact combination.
Retail $12.50
Rose Design on Straw lighter, compact and cuff links.
EC-54 Retail $15.00 Plus Tax
Also available in white. Specify.
Old Fashioned Carriage
EC-51 Retail $7.50
Flower Cart
EC-53 Retail $7.50
Also available in Chartreuse. Specify.
Butterfly
EC-52 Retail $7.50
Jewelbugs
EC-58 Retail $6.50
Zulu Head
EC-62 Retail $5.00
A Bluebird
EC-55 Retail $7.50
B Rose Design on Chartreuse
EC-60 Retail $5.00
Also available in white. Specify.
C Ruby Colored Enamel Stones
EC-59 Retail $7.50
Also available in multicolored stones. Specify.
D Stoned Butterfly
EC-57 Retail $6.50
E Redbirds
EC-56 Retail $6.50
F Petit Point Pattern on white
EC-61 Retail $5.00

Evans
ITEMS ATTRACTIVELY GIFT-BOXED
PRICES SUBJECT TO FEDERAL TAX
A fresh and ingenious new conception of costume jewelry in powdery pastel tones. Delicate contemporary classics in Evans Enamel. Decisive to the eye yet light and airy to the touch. Choice of straw, turquoise, or pastel green. Specify.
EC191 Necklace, earrings Retail $16.00
EC192 Necklace only 12.50
EC193 Earrings only 3.50
EC194 Lighter, link bracelet 18.50
EC195 Lighter, cuff bracelet 18.50
EC196 Lighter only 6.00
EC197 Bracelet only, link 12.50
EC198 Bracelet only, cuff 12.50
Trinkets of fashion finery created to appeal to the feminine need for vivid color accents. In luscious lipstick red Evans Enamel capped with 24K gold encrusting and beautifully complemented by the soft sheen of 14K gold electroplate. Also in royal blue or straw. Specify.
EC199 Necklace and earrings Retail $12.50
EC200 Necklace only 10.00
EC201 Earrings only 2.50
EC202 2 pieces (Lighter, Bracelet) 15.00
EC203 Lighter only 5.00
EC204 Bracelet only 10.00
Turquoise, golden straw and pastel green cuff bracelets are available as single pieces or matching sets of necklace, earrings, link bracelet and lighter as shown above.
EC155 Necklace and Earrings Retail $18.50
EC156 Necklace only 15.00
EC157 Earrings only 3.50
EC188 Lighter, Bracelet 21.00
EC189 Lighter only 6.00
EC190 Bracelet only 15.00

Evans
Cool green background provides proper setting for jewel-like topaz 3 dimensional Evans Enamel acorns on lighter and 4" ashtray.
EC227 2-piece set ... Retail $15.00
EC228 Lighter only 10.00
"Red Fire," the theme of this vividly conceived enamelware. Dancing multi-colored highlights surrounded by 24K gold encrusting create a warm glow that will fire the imagination. King sized cigarette box.
EC223 Lighter, cigarette box, 6" ashtray Retail $30.00
EC224 Lighter only 10.00
EC225 Compote 8" 22.50
EC226 Pair candlesticks 12.50
EC229 2-piece set Retail $15.00
EC230 Lighter only 10.00
A reflection of sunny vineyards, our "Vintage" pattern. Amethyst 3 dimensional Evans Enamel grapes complement delicate pastel green encrusted with 24K gold. King sized cigarette box.
EC233 Lighter, cig. box, 6" ashtray ..... Retail $31.00
EC234 Lighter only 12.50
EC235 8" compote 20.00

"EVENING CLUTCH." Exquisite pastel brocade in floral design with champagne satin lining. Pockets for every need.
EC340* .... Retail $17.50

"GALA." Emerald or black satin with contrasting applique of golden leaves and Jeweler's Bronze trim. Soft beige satin lining.
EC344* Specify choice .... Retail $18.00

"NOCTURNE." Black hand-buffed enamel, pink rose trimmed with golden finish metal. Satin lining, pockets. Also white or red (specify).
EC341* .... Retail $17.50

ELEGANCE
EVANS

EVANS

EVANS *proudly presents a distinguished and exclusive collection of new, original designs and fashion-styled personal accessories. . . . Those illustrated are in golden color metals, shimmering reflective highfired* EVANS *enamel, exquisite satin and brocade.... Again, we prove that* EVANS IS ELEGANCE.

"SERENADE." Radiant ruby and emerald quartz accented with diamond-like rhinestone rondels on highfired Evans enamel. Also available in sapphire. Specify.
EC345* Earrings .... Retail $3.50
EC346* Necklace ...... Retail $15.00
EC347* Bracelet .... Retail $15.00
EC348* Compact .... Retail $7.50
(Compact with exclusive top lift feature for proper release of powder.)
EC349* Mirror .... Retail $7.50
EC350* Necklace, earrings.. Ret. $18.50

"FLAIR." Vibrant red highfired enamel flecked with 24K gold. Also available in pink, white or black. Specify color.
EC351* Bracelet .... Retail $7.50
EC352* Earrings .... Retail $2.50
EC353* Necklace .... Retail $7.50
EC354* Necklace, earrings.. Ret. $10.00

"DAINTY SHAVER." Lovely accessory with dainty rose design on highfired white enamel to take full size double blade.
EC359 Retail $5.00

"DUAL MIRROR." A traveling accessory with both regular and magnifying mirrors for your convenience. Sturdy wire stand for propet balance. Swirl or basketweave design in Jeweler's Bronze. Specify style choice.
EC342 Swirl 5" .... Retail $7.50
EC343 Basketweave 4" .... Retail $5.00

"CELESTIAL." Gleaming rhinestone rondels and semi-precious sapphire quartz set on background of highfired Evans enamel. Delightful design accented with 24K gold. Also in ruby or emerald. Specify.

EC355* Lighter ...... Retail $7.50
EC356* Cigarette case.. Retail $7.50
EC357* Compact ..... Retail $8.50
(Compact with exclusive spring lever to hold your favorite pressed powder.)
EC358* Pill Box...... Retail $5.00
EC360* Lighter, case.. Retail $15.00

Evans

FRENCH ENAMEL (LIMOGES[illegible]

The "Jewelry of the Home"—Masterpieces of finest craftsmanship by the world's foremost creators of fashion-wise accessories in vitrified enamel fused to metal for beauty and permanence.

Jewel-embedded lighter, cigarette urn and "versa-tray"—An unfor[illegible] gettable gift for any occasion. Illustrated in Pastel Yellow with contrast[illegible] ing light Topaz enamel stones. Also available in Green, Olive an[illegible] Turquoise with contrasting enamel jewels. Gift packaged. Specify colo[illegible]
1A-367-2 .... Retail $42.5[illegible]

ALL VERSA TRAYS (candy dish or ash tray) FURNISHED WITH BALL FEET

Provincial Rooster struts proudly on this slim lighter. Pretty and decorative . . . In Yellow, Red and Gold on Jade Green background.
562-654-2..... Retail $10.00

Gay petit point pattern lighter is perfect for the person who loves the traditional, for it looks at home among cherished possessions. Design on white.
562-375-2.... Retail $10.00

Oo-la-la! Three-piece French poodle set with ash tray, lighter and cigarette urn. The adaptable "versa-tray" is a show place for the pert black poodle on a mottled gray background. Gift packaged.
1A-667-2 .... Retail $37.50

Slim jeweled cigarette box and table lighter make a set to be treasured. Case sits on four gold ball feet. Olive, Green, Straw or Turquoise with contrasting embedded enamel stones. Gift packaged. Specify color.
1K-367-2 .... Retail $35.00

ENTIRE BODY OF EACH LIGHTER IS FUEL RESERVOIR

Stylized stallion table lighter with matching ash tray. Perfect pair for office or den. Black stallion, gold trim on gray background. Gift packaged.
1B-369-2............... Retail $17.50

ALL ILLUSTRATIONS ½ ACTUAL S[illegible]

A gift he will long cherish—A smoker's set consisting of lighter, cigarette server and ash tray; handsome in golden metal and polished genuine leather. Mounted on 4 ball feet. Ash tray has removable glass inner-liner.
1R-27-2 Black Morocco
Retail $15.00
1R-27-2 Ostrich... Retail $17.50
1R-27-2 Alligator.. Retail $20.00

All Evans lighters have fiber glass lifetime wicks, red seal fuel control and patented filler which insures hundreds of extra lights.

Evans sensational, new "Petite" Carryall. Beautiful swirl design in two-tone gold. Metal mesh handle. Contains automatic lipstick, coin holder, automatic lighter, door-type powder compartment, velour puff, large make-up mirror and separate space for cigarettes.
H 42-131-17...... Retail $22.50
Plus 10% Tax

Evans "Lamp-Liter" for his office or den. A clever and useful gift: A fine lamp with 3-way socket and a bone-china table lighter in Onyx White, hand-decorated with game birds. Available with Green or Beige shades. Specify color.
LP-815-74-2 .... Retail $25.00
Game bird lighter only.
815-74-2 .... Retail $15.00

Evans

STYLE ACCESSORIES

DISTINCTIVE GIFTS FOR THE VERY SPECIAL PERSON IN YOUR LIFE

For the Man—A handsome pocket lighter or desk set. For the Woman—A dainty pocket lighter, a lovely set or one of the new and Oh!—So fashionable, glamorous Evans Carryalls.

Matching Compact and Lighter Set ornamented with a distinctive Heraldic design in Black and Gold. Petite compact has lup-lifter and double velour puff. Golden metal trim. Gift boxed.
EE-388-2 .... Retail $7.00

Evans Carryall. Graceful wave design in gleaming golden metal. Metal mesh handle. Contains automatic lipstick, coin holder, door-type powder compartment, velour puff, large make-up mirror and separate compartment for regular or king-sized cigarettes.
H 43-360-2...... Retail $27.50
Plus 10% Tax

Evans "Supreme" Men's Pocket Lighter. Tall and trim. Engine-turned design with engraving shield for initials.
51-25-2 Golden Metal. Retail $3.00
51-25-3 White Metal.. Retail $3.00

Evans "Pettie" Ladies Lighter—Engine-turned design, with engraving shield for initials.
55-619-2 Golden Metal. Ret. $3.00
55-619-3 White Metal.. Ret. $3.00

Evans Carryall. Distinctive moire design in two-tone gold. Chain handle. Contains automatic lipstick, coin holder, door-type powder compartment, velour puff, large make-up mirror and separate compartment for regular or king-size cigarettes.
HM 42-233-17.... Retail $30.00
Plus 10% Tax

Evans "Jumbo." A generously proportioned lighter for the man who wants months of carefree service. Extra large fuel capacity. Covered with genuine leather. Golden metal trim.
50-27-2 Light Brown Calf
Retail $ 5.00
50-27-2 Ostrich..... Retail $ 7.50
50-27-2 Alligator.... Retail $10.00

Evans "Banner." Graceful engine-turned design with engraving shield for initials.
52-20-2 Golden Metal. Retail $3.00
52-20-3 White Metal.. Retail $3.00

Evans Carryall. New engine-turned design decorated with simulated Sapphire mounts. Golden metal finish. Chain handle. Contains automatic lipstick, coin holder, door-type powder compartment, velour puff, large make-up mirror and separate compartment for regular and king-size cigarettes.
HM 43-234-2.... Retail $32.50
Plus 10% Tax

POCKET LIGHTERS ILLUSTRATED ½ ACTUAL SIZE.
CARRYALLS AND SMOKERS SET ILLUSTRATED ½ ACTUAL SIZE.

Evans

FINE WOMEN'S PERSONAL ACCESSORIES

Fashioned by the creators of famous Evans fitted handbags.

Evans "Serv-Paç" cigarette case in golden mesh. King size.
EC-37..............................Retail **$10.00**
Also available in "Regular."

Ladies Wallet in golden mesh.
EC-38.................Retail **$8.50** Plus Tax

Eyeglass Case in golden mesh.
EC-39.................Retail **$7.50** Plus Tax

Coin Purse in golden mesh.
EC-40.................Retail **$5.00** Plus Tax

All mesh items also available in white metal. Specify.

3 Piece Enam. Dresser Set
EC-43..............Retail **$15.00**
Also available in both Black and Straw. Specify. Gift Packaged.

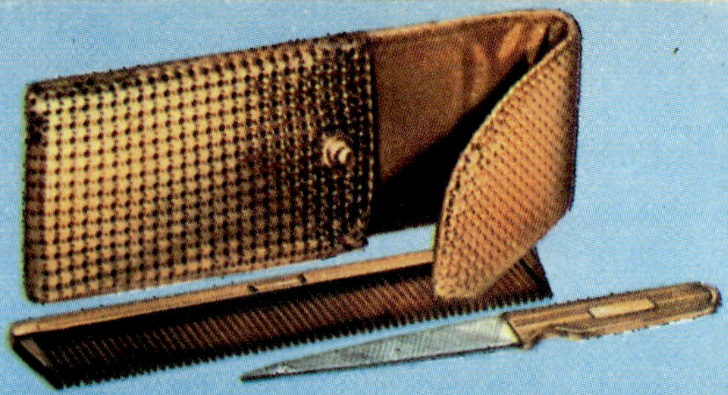

Comb and File Set in golden mesh.
EC-41..Retail **$7.50**

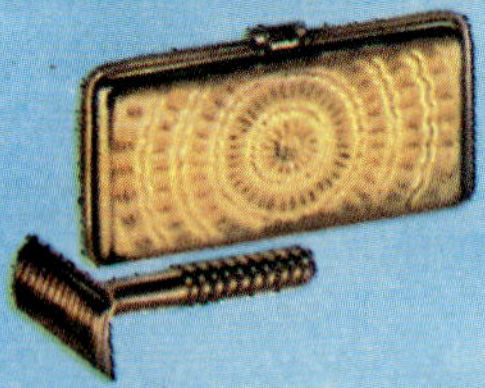

Miss Evans Little Shaver
EC-42......................................Retail **$1.50**

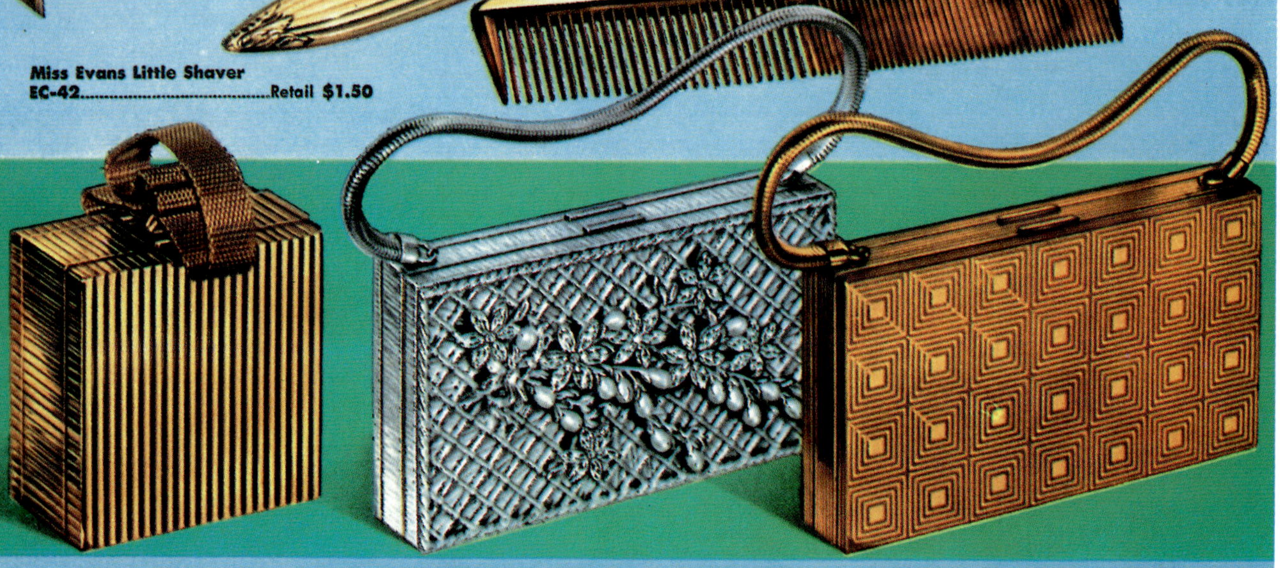

Golden Metal Carryall
EC-44...........................Retail **$17.50** Plus Tax

Rhinestone Decorated Carryall in rhodium
EC-45...........................Retail **$30.00** Plus Tax

Evans Carryall Engine Turned Design.
EC-46...........................Retail **$20.00** Plus Tax

All Evans carryalls contain automatic lipstick, coin holder, door type powder compartment, velour puff, make up mirror and compartment for cigarettes.

EVANS

Our new collection of smoking a[…] cessories in exclusive highfir[…] EVANS enamel has been design[…] for those who prefer the ultima[…] in styling.

The exclusive enamel process is t[…] result of the tireless efforts of t[…] EVANS scientists in research a[…] testing in order to obtain a p[…] fected secret formula.

Many of our works have been i[…] spired by the great artists of p[…] eras such as Cellini, Faberge, a[…] Leonardo da Vinci who stat[…] "Enamel is the eternal art! Enam[…] is the most lasting, most beauti[…] and most enduring form of dec[…] ration."

"BOUQUET." Central design in three dimensional hand-chased flowers highlighted with topaz enamel quartz trimmed with 24K gold. A beautiful harmony of color on Evans pastel straw highfired enamel. Hand-polished Jeweler's Bronze trim. 3¾" lighter, deep king-size cigarette box, 3½" diameter ashtrays.

EC328 Lighter. 4 trays, cig. box...Retail $32.50
EC329 Lighter, cigarette box......Retail $25.00
EC330 Bridge set: lighter, 4 trays..Retail $20.00
EC331 Lighter only ....Retail $12.50
EC332 Matching 8" compote or tray..Ret. $15.00

"ROUGE FLAMBE." A masterpiece of the enameling art. A dramatic smoking ensemble with throbs of color and verve. Multicolor highfired Evans enamel with rich hand-polished Jeweler's Bronze. 5" lighter, deep king-size cigarette box, and 6" ash tray.

EC323 3-piece set: lighter; ashtray; cigarette box.......Retail $30.00
EC324 Lighter and cigarette box only...............Retail $22.50
EC325 Lighter and ashtray only...................Retail $17.50
EC326 Lighter only ....Retail $10.00
EC327 Matching 8" compote or ashtray..............Retail $15.00

"HARVEST." Colorful fruit in repousse on Florentine green bac[…] ground. An artistic creation in highfired Evans enamel and trimm[…] with hand-polished Jeweler's Bronze. Lighter and small ashtrays a[…] ideal for your guests' convenience at the bridge table. 3¾" light[…] 3½" ashtrays, deep king-size cigarette box.

EC333 Lighter, 4 ashtrays, cigarette box..............Retail $32.[…]
EC334 Lighter and cigarette box ....Retail $25.[…]
EC335 Bridge set: lighter and 4 ashtrays............Retail $20.[…]
EC336 Lighter only ....Retail $12.[…]
EC337 Matching 8" compote or ashtray..............Retail $15.[…]

"IMPERIAL EGG." Collector's item (with Faberge touch). Opens to reveal artfully concealed cigarette lighter. Delicate highfired Evans enamel trimmed with pierced gallery in Antique Finish. Matching ashtray. Available in pastel straw, pink, blue, or white. Specify color.

EC338 Lighter only ....Retail $15.00
EC339 Set: lighter and ashtray........Retail $20.00

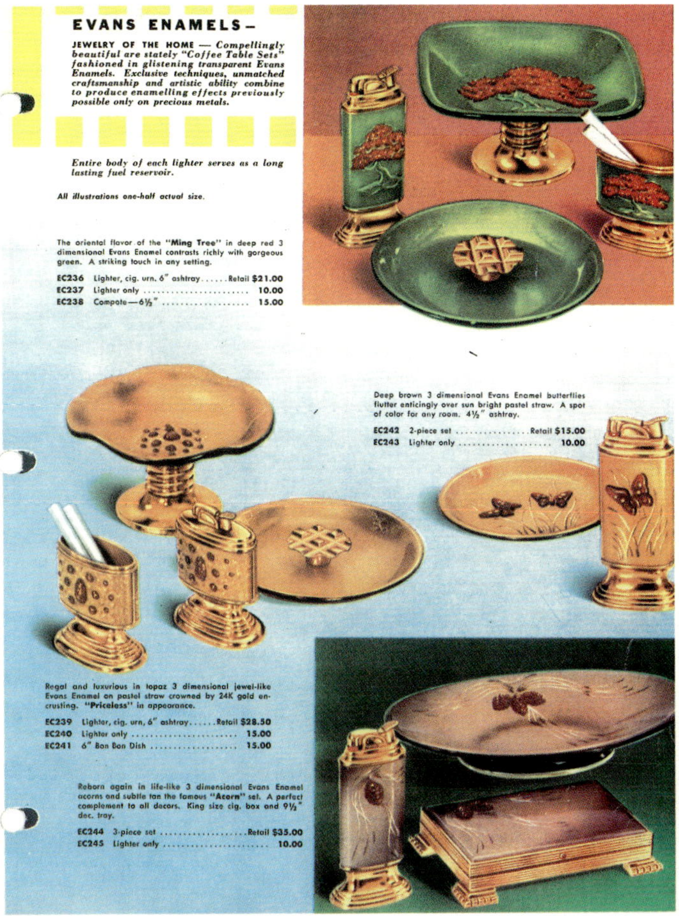

EVANS ENAMELS–

JEWELRY OF THE HOME — *Compellingly beautiful are stately "Coffee Table Sets" fashioned in glistening transparent Evans Enamels. Exclusive techniques, unmatched craftsmanship and artistic ability combine to produce enamelling effects previously possible only on precious metals.*

Entire body of each lighter serves as a long lasting fuel reservoir.

All illustrations one-half actual size.

The oriental flavor of the **"Ming Tree"** in deep red 3 dimensional Evans Enamel contrasts richly with gorgeous green. A striking touch in any setting.

EC236 Lighter, cig. urn. 6" ashtray......Retail $21.00
EC237 Lighter only 10.00
EC238 Compote—6½" 15.00

Deep brown 3 dimensional Evans Enamel butterflies flutter enticingly over sun bright pastel straw. A spot of color for any room. 4½" ashtray.

EC242 2-piece set ....Retail $15.00
EC243 Lighter only 10.00

Regal and luxurious in topaz 3 dimensional jewel-like Evans Enamel on pastel straw crowned by 24K gold encrusting. **"Priceless"** in appearance.

EC239 Lighter, cig. urn, 6" ashtray......Retail $28.50
EC240 Lighter only 15.00
EC241 6" Bon Bon Dish 15.00

Reborn again in life-like 3 dimensional Evans Enamel acorns and subtle tan the famous **"Acorn"** set. A perfect complement to all decors. King size cig. box and 9½" dec. tray.

EC244 3-piece set ....Retail $35.00
EC245 Lighter only 10.00

Evans®

EVANS GENUINE CLOISONNÉ ENAMEL

THE AGELESS ART OF ENAMELLING BROUGHT TO MODERN DAY PERFECTION BY EVANS. Exclusive handcrafted designs fired at 1450° Fahrenheit for beauty and permanence.

Entire body of each lighter is long lasting fuel reservoir.

Contemporary three piece smoking set in unique Universe pattern. Set features Evans automatic lighter, cigarette box and 8" decorator tray in genuine cloisonné enamel.

EC-110 3 piece set....Retail $35.00
EC-111 Lighter only....Retail $10.00

Ever popular dainty rosettes grace this charming three piece Evans genuine cloisonné enamel smoking set in the traditional manner. Also available in black. Specify.

EC-114 3 piece set....Retail $25.00
EC-115 Lighter Only....Retail $12.50

Pine Cone decorator tray, cigarette box and automatic lighter in beautiful golden rust Evans genuine cloisonné enamel.

EC-117 3 piece set....Retail $30.00
EC-118 Lighter only....Retail $10.00

For that current Oriental trend in home decor—this rich olive green two piece Evans genuine cloisonné enamel smoking set decorated with deep red Ming Tree design in raised enamel.

EC-112 2 piece set....Retail $22.50
EC-113 Lighter only..Retail $10.00

Distinctive COAT OF ARMS pattern adorns regal Evans automatic lighter and ash tray of two piece smoking set in Evans genuine cloisonné enamel.

EC-116 2 piece set....Retail $15.00

All sets attractively gift packaged. Illustrations approximately one-half actual size.

Evans®

The fabulously popular **"Hitching Post"** Evans Fine China smoking set. Handsome novel stallion lighter complementing king size cigarette box and 5½" ashtray, all in soft eggshell black. Also in white, specify.

EC246 3-piece set ....Retail $20.00
EC247 Lighter only ....Retail $12.50

The **"Lamplighter"** original Evans Creation. Graceful lamp serves as base for jewel-like decorated classical Fine China lighter. Lamp features three way socket.

EC249 Lamp and Lighter ....Retail $30.00
EC250 Lighter only 20.00

The **"Butler"** unique cigarette lighter-ashtray combination. Eggshell beige finish contrasts prettily with delicate gold touched leaves. Lighter removes for filling and cleaning. Also in black, specify.

EC248.............................Retail $12.50

In tune with the times, this Evans Fine China smoking set centering around the majestic **"Chessman"** table lighter, 5½" ashtray and king size cigarette box. Palomino beige shown. Also in white, specify.

EC251 3-piece set ....Retail $25.00
EC252 Lighter only 15.00

Finely sculptured dancing figures adorn 4" ashtrays and **"Stein"** lighter of this Evans Fine China smoking set styled in the old world bavarian tradition. At home in any room.

EC253 3-piece set..................Retail $15.00
EC254 Lighter only 10.00

Evans
LIGHTERS AND SMOKING SETS IN LASTING EVANS ENAMEL
. . . style and color-toned for modern living.
. . . gracious luxury with a fashion flair.
. . . gifts with a purpose and a place in any home.
ENTIRE BODY OF EACH LIGHTER IS A FUEL RESERVOIR
ALL SETS ATTRACTIVELY GIFT PACKAGED.
Pine Cone 3 Piece Set
18EC2100 Retail $35.00
Lighter Only
19EC600 Retail $10.00
Poodle
21EC600 Retail $10.00
Rooster
20EC600 Retail $10.00
Flower Cart 2 Piece Set
22EC1350 Retail $22.50
Lighter Only
23EC600 Retail $10.00
Old Fashioned Carriage 2 Piece Set
24EC900 Retail $15.00
Lighter Only
25EC600 Retail $10.00
Vintage 2 Piece Set
26EC1350 Retail $22.50
Lighter Only
27EC600 Retail $10.00

Evans
FINE BONE CHINA
TABLE LIGHTERS AND SMOKING SETS
Decorator-styled in Precious Gem Colors
Distinctively new Table Lighter. Hand-decorated rose on Onyx White.
821-31-2
Retail.......$12.50 No Tax
4-Piece Smoker's Set. Contemporary design in Basalt Black. Hand-decorated in bright Gold. Also available in Onyx White or Yellow. Gift packaged. Specify color.
81A-4-2.......Retail $12.50 No Tax
Glamorous Jeweled Table Lighter in Antiqued Ivory with enamel jewels. Hand-decorated with Gold.
817-50-2
Retail...$20.00 No Tax
3-Piece Smoker's Set. Exquisitely designed in the modern manner. Illustrated in Lapis Blue. Also in Onyx White, Pearl Gray, Jade Green or Basalt Black. (See color samples at left.) Gift packaged. Specify color.
84F-6-2.......Retail $15.00 No Tax
Classic Urn Table Lighter in Jade Green with hand-decorated leaf pattern in bright Gold. Also available in Onyx White. Specify color.
814-55-2
Retail......$15.00 No Tax
ENTIRE BOWL IS FUEL RESERVOIR INSURING A YEAR'S SUPPLY OF LIGHTS
Matching Ashtray in Onyx White, hand-trimmed in Gold. Also in Lapis Blue. Specify color.
820-2-2.....Retail $3.00 No Tax
Modern Urn-Shaped Table Lighter in Pearl Gray. Also in Jade Green. Specify color.
815-1-2.........Retail $15.00 No Tax
Melon Pattern Table Lighter in Onyx White, hand-trimmed in Gold. Also in Lapis Blue. Specify color.
816-2-2.......Retail $15.00 No Tax
ALL ILLUSTRATIONS ½ ACTUAL SIZE

Evans
EVANS ACCESSORIES FOR 1957
Fascinating new and novel accessories by Evans, the Fashion Leaders.
More firepower for '57 with fully automatic pocket lighters and brilliant new "FLIPTOP" windlighters featuring exclusive "CLEAN SPINNING" lighting action.
A flourish for mixing techniques. Distinctive bar set in "Jewelers Bronze" and fiery red. Includes jigger, spoon, fork, and can-bottle opener in gift package. Also in black, specify.
EC264.......Retail $15.00
A. "Petite" ladies automatic lighter in engine turned design. A polished chrome "Jewel." Also in golden "Jewelers Bronze," specify.
EC265.......Retail $3.00
B. The popular "Banner" automatic lighter featuring "Red Seal" fuel control. In "Jewelers Bronze," engine turned design, also in polished chrome, specify.
EC266.......Retail $3.00
C. Evans automatic lighter featuring relief "Saint Christopher" medal.
EC267.......Retail $3.50
D. Thin "Fliptop" windlighter with clean spinning action in shimmering chrome, engine turned design. Also in golden "Jewelers Bronze," specify.
EC268.......Retail $4.00
E. Generous sized "Fliptop" windlighter gives many lights per filling. Navy lizard vinyl and polished chrome. Also in black, specify.
EC269.......Retail $4.00
F. Vivid red touch of masculine color decorates this "Fliptop" windlighter.
EC270.......Retail $5.00
G. "Fliptop" windlighter, tall and trim in glistening "Jewelers Bronze," engine turned design. Also in polished chrome, specify.
EC271.......Retail $5.00
H. Surefire "Fliptop" windlighter, polished chrome with accent of jet black Evans Enamel. Also in golden "Jewelers Bronze", specify.
EC272.......Retail $6.00
This night owl holds up a "light" for many hours of whimsical smoking pleasure. The "Gay boy" in soft black eggshell Evans Fine China. Removable lighter unit.
EC273.......Retail $15.00
An all-time favorite man's gift item, the "Executive" set in simulated black morocco and "Jewelers Bronze." Removable automatic lighter unit and fine china ashtray. Also in brown, specify.
EC274.......Retail $10.00
Slim gold lines set off the jet black Evans Enamel lending distinction to "Fliptop" windlighter, cuff links and tie bar of exclusive gift packaged men's jewelry set. Cuff links feature easy snap apart action.
EC275.......Retail $12.50 Plus $ .50 Tax
tho in U.S.A.

"THE SPOOL." Slim tapered elegance keynotes this Evans Fine China set trimmed with gleaming 24K gold fish fired on lustrous black glaze, swirled with gray overglaze. 6¾" lighter, urn, 5½" ashtray.
EC300 3-piece set.......Retail $17.50
EC301 Lighter only.......Retail $12.50
"SILHOUETTE." Fascinating echo of ancient cultures brought to modern beauty in sleek contour set designed by Sam Speers. Black stippled with 24K gold on Evans Fine China. 8" lighter, urn, 5½" ashtray.
EC302 3-piece set.......Retail $22.50
EC303 Lighter only.......Retail $15.00
ELEGANCE
EVANS
EVANS
For the past half century, the EVANS skillful artisans, under the guidance of A. F. Reilly, Sr., founder of the firm, have produced the leading fashion accessories for Home and Personal use. . . . This year's collection includes interpretations of many classical works of art, which have been created and manufactured at our plant.
"CLASSICAL." Attractive ensemble. Ideal for card table use. In Evans Fine China encrusted with 24K gold. Set includes lighter, urn and 2 ashtrays, excellent for entertaining. Specify color: black or white.
EC308 4-piece set.......Retail $12.50
"RENAISSANCE." Three-piece set in Evans Fine China. Transparent glaze in soft, warm earthen tones with contrasting dark brown overglaze. Enriched with emerald-like quartz and 24K gold accents.
EC361 3-piece set.......Retail $25.00
EC361A Lighter Only.......Retail $15.00
"CONTEMPORARY." Refreshingly new in design, a three-piece smoking set with the famous Sure-Fire Evans Lighter. Glossy black, turquoise or ivory glaze on Evans Fine China with highlights of 24K gold. Specify color: black, turquoise or ivory.
EC362 3-piece set.......Retail $12.50
EC363 Lighter only.......Retail $7.50
"CHUM." Aristocratic black or white poodle, hand decorated and finished in Evans Fine China with golden finish collar and chain. Poodle stands 6¼" high on circular black glass tray, 2½" lighter and ashtray on either side. Specify color: black or white.
EC309 Poodle set.......Retail $15.00

EVANS

"EXECUTIVE SET." Impressive and useful accessory for home or office. Rectangular shape trimmed with simulated black or brown morocco with fine polished Jeweler's Bronze top and bottom. **Specify color.**
EC312 .... Retail **$10.00**

"HITCHING POST." All the beauty of a champion stallion expertly captured in this skillfully modelled gentleman's smoking set. Evans Fine China, in black or white. 7¼" lighter with matching urn and 5½" ashtray. **Specify choice of color.**
EC310 3-piece set .... Retail **$17.50**
EC311 Lighter only .... Retail **$12.50**

"TALLY-HO." All the excitement of the hunt is symbolized in this set of Evans Fine China in black or brown. Carefully detailed 6½" riding boot lighter, urn and 5½" ashtray. **Specify choice of color.**
EC364 3-piece set .... Retail **$15.00**
EC365 Lighter only .... Retail **$10.00**

"COCKTAIL HOUR" BAR SET. All the required gadgets to be an expert mixologist for your guests. Evans Enamel handles in shaded flame or black with Jeweler's Bronze implements. Gift-boxed. **Specify color.**
EC316 .... Retail **$15.00**

"SPOOL" BAR SET. In keeping with the fashion trends, this four-piece bar set matches the Spool lighter set. Evans Fine China handles top the Jeweler's Bronze tools. Available 3 ways.
EC366 Boxed Bar Set .... Retail **$20.00**
EC367 Bar set and stand .... Retail **$35.00**
EC368 Stand only .... Retail **$75.00**

"GAY BLADE." A whimsical set in which golden moonlight shines down on late reveler holding up a lamppost lighter. Gold highlights on black Evans Fine China with removable lighter. Matching urn 9¼" tall.
EC322 .... Retail **$15.00**

"ANTIQUE AUTO SET." Unique three-piece smoking set in gleaming white Evans Fine China. Stein lighter, two matching ashtrays decorated with antique auto motif in soft tones highlighted with 24K gold trim.
EC369 3-piece set.... Retail **$12.50**
EC369A Lighter only.. Retail **$10.00**

"IMPERIAL EGG." Collector's item (with Faberge touch). Opens to reveal artfully concealed cigarette lighter. Delicate highfired Evans enamel trimmed with pierced gallery in Antique Finish. Matching ashtray. In pastel straw, pink, blue or white. **Specify color.**
EC338 Lighter only .... Retail **$15.00**
EC339 Set: Lighter and Ashtray .... Retail **$20.00**

JEWELED ENSEMBLE

IN GOLDEN METAL WITH SAPPHIRE RHINESTONES

TOP – CARRYALL No. HM43-358-2 RETAIL $35.00 PLUS TAX
With space for cigarettes, coins, lipstick and powder . . . has full size mirror, comb and velour puff.

LEFT – POWDER BOX	No. 459-358-2	RETAIL $12.50
LIPSTICK HOLDER	No. 94-358-2	RETAIL 10.00
LIGHTER	No. 5-358-2	RETAIL 10.00
SERV-PAC (Armor Mesh, Holds full pack, king size)	No. 28-358-2	RETAIL 17.50 ALL PLUS TAX

®Evans

7

4-pc. set illustrated in White

Also furnished in Yellow or Black. Specify color

No. 81A-4-2 RETAIL $12.50

GIFT BOXED

TABLE LIGHTER ONLY

No. 81-4-2 RETAIL $7.50

FINE BONE CHINA
HAND-DECORATED
WITH BRIGHT GOLD

®Evans

4

FINE HAND-DECORATED BONE CHINA TABLE LIGHTERS

TOP . .	No. 814-31-2	5″ high	RETAIL	$17.50
LEFT . .	No. 817-31-2	4″ high	RETAIL	17.50
RIGHT . .	No. 815-31-2	4½″ high	RETAIL	17.50
ASHTRAY .	No. 822-31-2	4½″ diam.	RETAIL	5.00

ENTIRE BOWL IS FUEL RESERVOIR — GIVES LIGHTS —
NOT FOR MONTHS — BUT FOR YEARS PER FILLING

® Evans

AUTOMATIC TABLE LIGHTERS

THE VICTORIAN*
555-1-2 — Golden metal finish.
555-1-22 — In silver electroplate.

THE DUCHESS*
556-7-22 — Silver electroplate.
556-7-2 — In golden metal finish.

THE FUTURA*
62-179-22 — Cigarette holder in silver electroplate; matches **554-179-22** table lighter.
554-179-22 — Silver electroplate.

THE NEW MAGIC LAMP*
550-176-22 — Silver electroplate.
550-176-2 — In golden metal finish.

THE STARDUST
555-4-2 — Golden metal finish.

THE WINDSOR
555-156-22 — Silver electroplate.
555-156-2 — In golden metal finish.

THE IMPERIAL
558-171-2 — Golden metal finish.
558-171-22 — In silver electroplate.

THE FAMOUS MAGIC LAMP
550-26-2 — Golden metal finish.
550-26-22 — In silver electroplate.

*Prices do not include engraving.

Evans

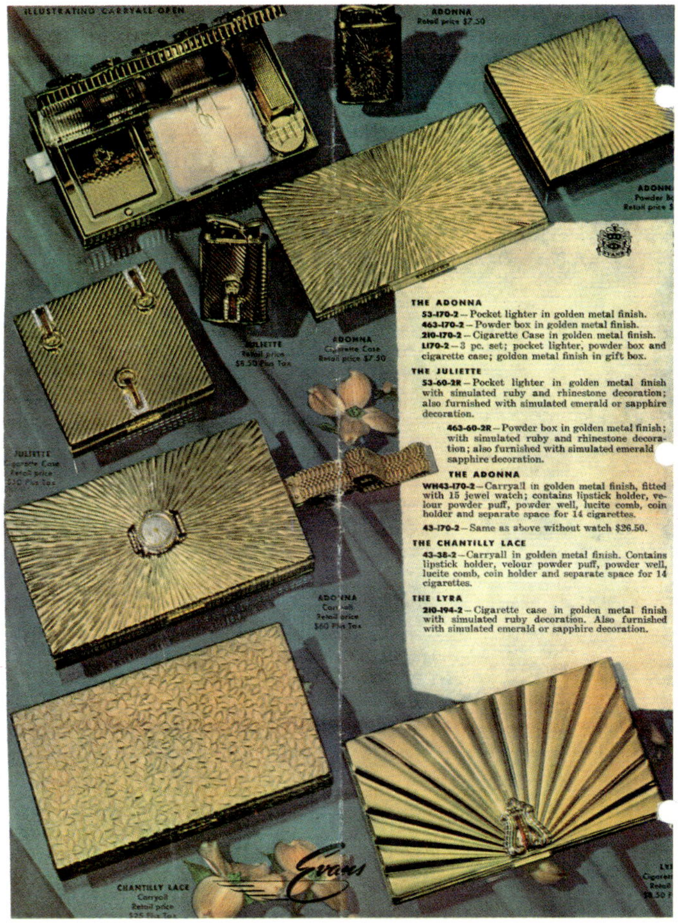

THE ADONNA
53-170-2 – Pocket lighter in golden metal finish.
463-170-2 – Powder box in golden metal finish.
210-170-2 – Cigarette Case in golden metal finish.
L170-2 – 3 pc. set; pocket lighter, powder box and cigarette case; golden metal finish in gift box.

THE JULIETTE
53-60-2R – Pocket lighter in golden metal finish with simulated ruby and rhinestone decoration; also furnished with simulated emerald or sapphire decoration.
463-60-2R – Powder box in golden metal finish; with simulated ruby and rhinestone decoration; also furnished with simulated emerald sapphire decoration.

THE ADONNA
WH43-170-2 – Carryall in golden metal finish, fitted with 15 jewel watch; contains lipstick holder, velour powder puff, powder well, lucite comb, coin holder and separate space for 14 cigarettes.
43-170-2 – Same as above without watch $26.50.

THE CHANTILLY LACE
43-38-2 – Carryall in golden metal finish. Contains lipstick holder, velour powder puff, powder well, lucite comb, coin holder and separate space for 14 cigarettes.

THE LYRA
210-194-2 – Cigarette case in golden metal finish with simulated ruby decoration. Also furnished with simulated emerald or sapphire decoration.

Evans

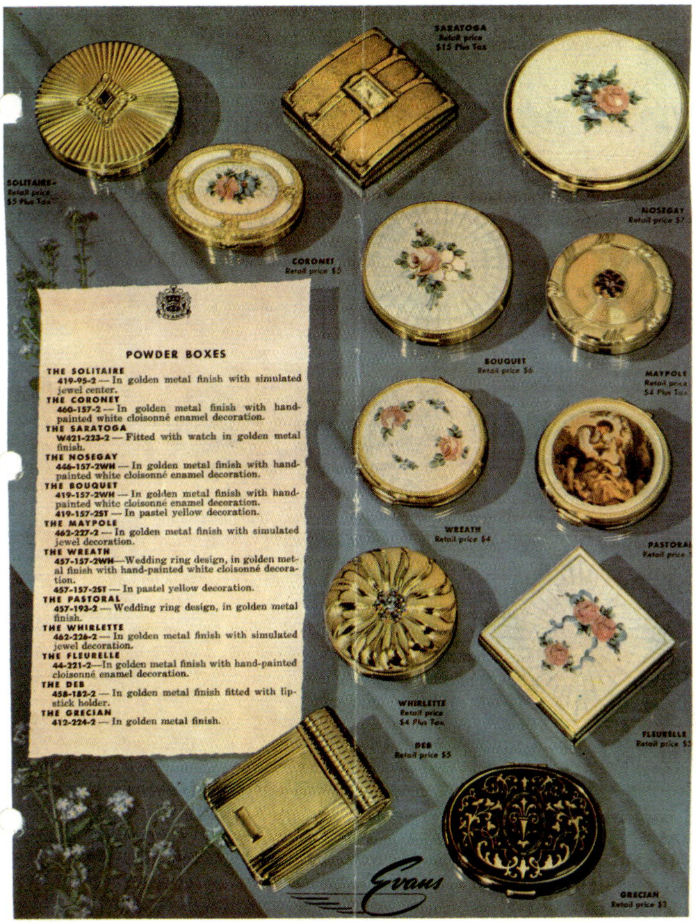

POWDER BOXES

THE SOLITAIRE
419-95-2 — In golden metal finish with simulated jewel center.
THE CORONET
460-157-2 — In golden metal finish with hand-painted white cloisonné enamel decoration.
THE SARATOGA
W421-223-2 — Fitted with watch in golden metal finish.
THE NOSEGAY
446-157-2WH — In golden metal finish with hand-painted white cloisonné enamel decoration.
THE BOUQUET
419-157-2WH — In golden metal finish with hand-painted white cloisonné enamel decoration.
419-157-2ST — In pastel yellow decoration.
THE MAYPOLE
462-227-2 — In golden metal finish with simulated jewel decoration.
THE WREATH
457-157-2WH — Wedding ring design, in golden metal finish with hand-painted white cloisonné decoration.
457-157-2ST — In pastel yellow decoration.
THE PASTORAL
457-192-2 — Wedding ring design, in golden metal finish.
THE WHIRLETTE
462-226-2 — In golden metal finish with simulated jewel decoration.
THE FLEURELLE
44-221-2—In golden metal finish with hand-painted cloisonné enamel decoration.
THE DEB
458-182-2 — In golden metal finish fitted with lipstick holder.
THE GRECIAN
412-224-2 — In golden metal finish.

Evans

THE 20-PAK...33-25-3—Combination cigarette case and lighter, non-tarnishing rhodium finish, holds regular or king size cigarettes . . . **33-25-2**—In golden metal finish.
THE WARWICK...52-87-2—Automatic pocket lighter in golden metal finish . . . **52-87-1**—In nickel finish **$3.00**
THE BARON . . . 31-177-1—Combination cigarette case and lighter in nickel finish, holds 14 cigarettes . . . **31-177-2** – In golden metal finish **$6.00**
THE STARFLOWER . . . 34-107-2—Combination cigarette case and lighter in golden metal finish, holds 10 cigarettes...**34-107-3** —In non-tarnishing rhodium finish.
THE BARON . . . 52-177-1—Automatic pocket lighter, in nickel finish, matches cigarette case **224-177-1 . . . 52-177-2**—In golden metal finish, matches cigarette case **224-177-2**—**$4.00**
THE KINGSTON . . . 56-46-2—Automatic pocket lighter—in golden metal finish.
EMBLEMS . . . 52-160-2—Automatic pocket lighter in golden metal finish, also furnished with Masonic, K of C, Eagles, L.O.O.M., Kiwanis, Rotary, Odd Fellows, Shriners, Lions and 32nd degree Masonic.
THE SPENCER . . . 52-22-2—Automatic pocket lighter in golden metal finish, matches cigarette case **26-22-2**.
THE BARON . . . 224-177-1—Cigarette case, nickel finish, holds full pack, matches lighter **52-177-1 . . . 224-177-2**—In golden metal finish, matches lighter **52-177-2—$3.00**
THE MONTEREY...31-178-2—Combination cigarette case and lighter in golden metal finish . . . **31-178-1**—In nickel finish **$5.00**
THE COSMOPOLITAN . . . 51-25-1—Automatic pocket lighter in nickel finish . . . **51-25-2**—In golden bronze finish **$5.00**
THE CHANTILLY LACE . . . 56-38-2—Automatic pocket lighter in golden metal finish.
THE SPENCER . . . 26-22-2—Cigarette case in golden metal finish, matches pocket lighter **52-22-2**.
THE MARQUIS . . . 51-188-2—Automatic pocket lighter in golden metal finish . . . **51-188-1**—In nickel finish **$5.00**
THE KENMORE...31-153-1—Combination cigarette case and lighter, nickel finish, holds 14 cigarettes . . . **31-153-2**—In golden metal finish **$6.00**

Evans

IMPERIAL EGG by

Evans

Table lighter cleverly concealed within a Faberge inspired egg. Evans highfired enamel combined with Renaissance metal treatment in a lovely Antique Finish.

EVANS CASE CO.

North Attleboro Massachusetts

A Upright Egg lighter — 52/522

B Upright Rosary Case w/Cross — 55/517
Horizontal Rosary Case w/Cross — 55/555

C Small Upright Perfume Holder — 55/527
Large Upright Perfume Holder — 55/512

D Horizontal Egg Cigarette lighter — 52/565
Horizontal Egg Ashtray — 53/227

E Upright Egg Cigarette lighter — 57/537
Round Ashtray — 53/69
Cigarette Urn — 54/614

*All items above available in all finishes: Golden and Silver Satin — White, Pink, Yellow and Blue Enamel

Evans

DUCHESS
Automatic table or desk lighter; new model; holds long lasting fuel supply. Satin finish silver or gold electroplate.
No. 556/7/22 silver finish Retail $10.00
No. 556/[illegible]2 gold finish Retail $10.00

MAGIC LAMP
Automatic table or desk lighter; large fuel capacity. Satin finish silver or gold electroplate.
No. 550/26/22 silver finish Retail $10.00
No. 550/26/2 gold finish Retail $10.00

WINDSOR
Automatic table or desk lighter, beautiful chased design in gleaming silver or gold electroplate.
No. 555/156/22 silver finish Retail $8.00
No. 555/156/2 gold finish Retail $8.00

SILVER QUEEN
Automatic table or desk lighter in satin finish silver or gold electroplate.
No. 555/6/22 silver finish Retail $7.50
No. 555/6/2 gold finish Retail $7.50

SILVER KNIGHTS
"Twin" table lighters in silver or gold electroplate, long lasting fuel capacity. In velvet lined gift box.
Set No. C14 silver finish, boxed Retail $12.50
Set No. C15 gold finish, boxed Retail $12.50
Individual Lighter No. 554/84/22 silver finish Retail $6.00
Individual Lighter No. 554/84/2 gold finish Retail $6.00

PRINCESS
Automatic table or desk lighter in satin finish silver or gold electroplate.
No. 553/59/22 silver finish Retail $5.00
No. 553/59/2 gold finish Retail $5.00

LONGFELLOW
Combination cigarette case and lighter; thin model; deep cut engine-turned design front and back, in gold electroplate. Holds 13 cigarettes.
No. 38/30/2 Retail $10.00

EMPRESS
Combination cigarette case and lighter; quilted design front and back, in gold electroplate. Holds 11 cigarettes.
No. 34/128/2 Retail $10.00

CAVALIER
Combination cigarette case and lighter, deep cut engine-turned design in nickel finish with black enamel stripes. Holds 11 cigarettes.
No. 36/24/1 Retail $7.50

CONTINENTAL
Combination cigarette case and lighter, thin model; deep cut engine-turned design front and back, in gold or nickel finish.
No. 38/43/2 gold finish Retail $10.00
No. 38/43/1 nickel finish Retail $10.00

70/005
10.00 pr.
70/007
15.00 pr.
70/006
10.00 pr.
70/001
24.50 pr.
52/548
20.00
70/009
25.00 pr.
53/515
15.00
52/549
25.00
55/706
5"/12.50
55/708
6"/15.00
99/513
10.00 set
53/514
7"/12.50
53/513 5"/10.00
entertaining accessories & NEW FALL ADDITIONS

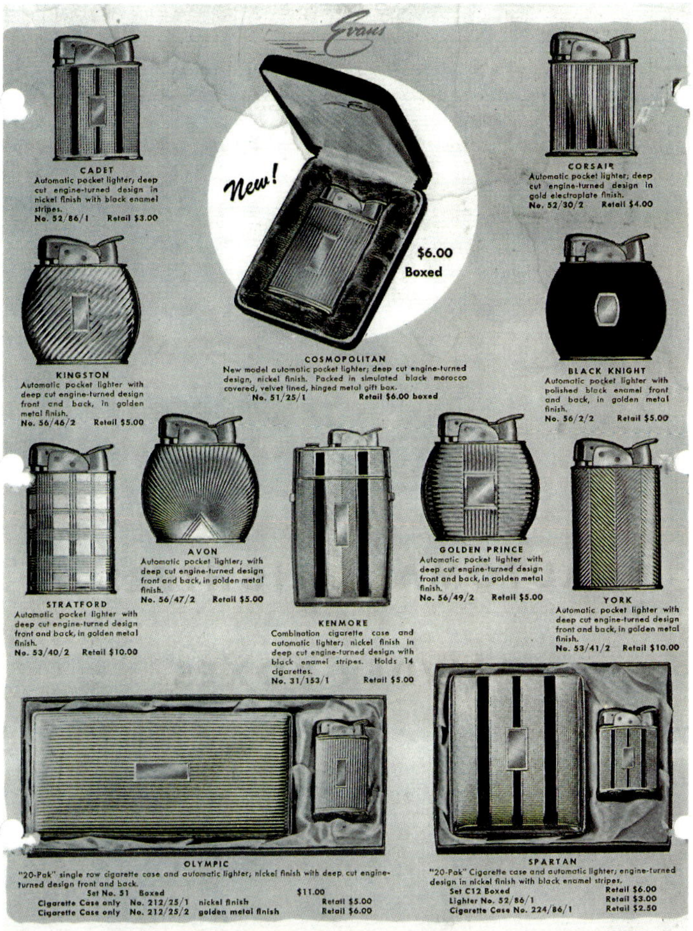
Evans
New!
$6.00
Boxed
CADET
Automatic pocket lighter; deep cut engine-turned design in nickel finish with black enamel stripes.
No. 52/86/1 Retail $3.00
CORSAIR
Automatic pocket lighter; deep cut engine-turned design in gold electroplate finish.
No. 52/30/2 Retail $4.00
KINGSTON
Automatic pocket lighter with deep cut engine-turned design front and back, in golden metal finish.
No. 56/46/2 Retail $5.00
COSMOPOLITAN
New model automatic pocket lighter; deep cut engine-turned design, nickel finish. Packed in simulated black morocco covered, velvet lined, hinged metal gift box.
No. 51/25/1 Retail $6.00 boxed
BLACK KNIGHT
Automatic pocket lighter with polished black enamel front and back, in golden metal finish.
No. 56/2/2 Retail $5.00
STRATFORD
Automatic pocket lighter with deep cut engine-turned design front and back, in golden metal finish.
No. 53/40/2 Retail $10.00
AVON
Automatic pocket lighter; with deep cut engine-turned design front and back, in golden metal finish.
No. 56/47/2 Retail $5.00
KENMORE
Combination cigarette case and automatic lighter; nickel finish in deep cut engine-turned design with black enamel stripes. Holds 14 cigarettes.
No. 31/153/1 Retail $5.00
GOLDEN PRINCE
Automatic pocket lighter with deep cut engine-turned design front and back, in golden metal finish.
No. 56/49/2 Retail $5.00
YORK
Automatic pocket lighter with deep cut engine-turned design front and back, in golden metal finish.
No. 53/41/2 Retail $10.00
OLYMPIC
"20-Pak" single row cigarette case and automatic lighter; nickel finish with deep cut engine-turned design front and back.
Set No. 51 Boxed $11.00
Cigarette Case only No. 212/25/1 nickel finish Retail $5.00
Cigarette Case only No. 212/25/2 golden metal finish Retail $6.00
SPARTAN
"20-Pak" Cigarette case and automatic lighter; engine-turned design in nickel finish with black enamel stripes.
Set C12 Boxed Retail $6.00
Lighter No. 52/86/1 Retail $3.00
Cigarette Case No. 224/86/1 Retail $2.50

ITEMS ON THIS PAGE FREE FROM RETAIL FED. EXCISE TAX
PRICES SHOWN ARE FAIR TRADE RETAIL
EVANS SUPREME
Tall and trim — the Evans Supreme is smart in appearance and holds enough fuel for weeks of constant use.
EVANS SUPREME
White metal—Engine turned design with Windguard
WG51-415-3 Retail: $3.75
EVANS SUPREME
Golden Metal—Engine-turned design
51-421-2 Retail: $3.95
EVANS SUPREME
White Metal—Butler and Polished
51-188-3 Retail: $3.95
ILLUSTRATIONS
ABOUT ¾ ACTUAL SIZE
EVANS SUPREME
White metal—Engine turned design
51-415-3 Retail: $3.50
Golden Metal
51-415-2 Retail: $3.50
EVANS SUPREME
White Metal—Engine-turned design
51-25-3 Retail: $3.50
Golden Metal
51-25-2 Retail: $3.50
EVANS SUPREME
White Metal—Brown Alligator Calf
51-416-3 Retail: $3.95
Golden Metal
51-416-2 Retail: $3.95
THERE IS NO CHARGE FOR SERVICING EVANS LIGHTERS
Evans
PLATE 2
2

EVANS
ILLUSTRATIONS ABOUT ¾ ACTUAL SIZE • PRICES FAIR TRADE RETAIL
EVANS BANNER
White metal — Engine turned design
Black enamel stripe
52-53-3 . . . . Retail $3.00
Golden Metal
52-53-2 . . . . Retail $3.00
EVANS KING
White metal — Engine turned design
Black enamel stripes
51-619-3 . . . . Retail $3.00
Golden Metal
51-619-2 . . . . Retail $3.00
EVANS BANNER
Golden Metal
French Enamel (Limoges)
Roses on White
52-157-2 . . . . Retail $4.00
NEW FIREBALL WICK
EVANS BANNER
White metal — St. Christopher
52-611-3 . . . . Retail $3.50
EVANS KING
White metal — Engine turned design
Black enamel — with windguard
and new "Fireball" Wick
WG51-415-3 . . . Retail $3.50
EVANS BANNER
Golden metal — Tortoise enamel
52-188-2 Tortoise . . Retail $3.50
Plain white metal — Butler and polished
52-188-3 . . . . . Retail $3.50
EVANS PETITE
Golden metal — Engine turned
55-619-2 . . . . Retail $3.00
EVANS BANNER
Golden metal — Ocean Pearl front
52-229-2 . . . . Retail $5.00
EVANS PETITE
Golden metal — Simulated leather front and back
55-10-2 alligator . Retail $3.00
Also available in simulated snake and ostrich.
Evans
AMERICA'S BEST AUTOMATIC LIGHTER
Smart new designs — by the same craftsmen who create the famous Evans fitted handbags, powder boxes, cigarette cases, and other fine accessories.
PLATE 1
1

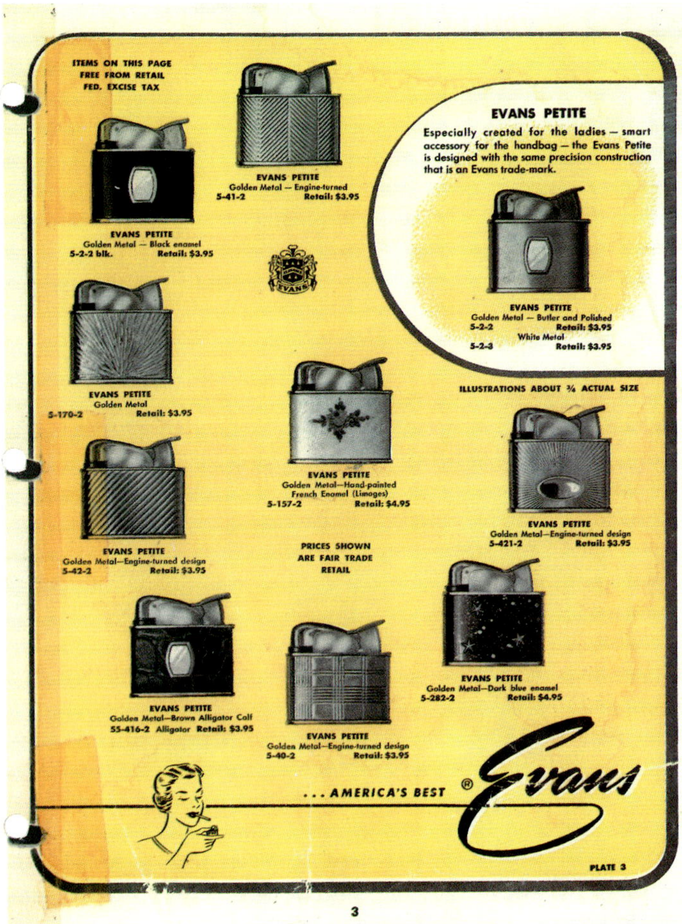

ITEMS ON THIS PAGE FREE FROM RETAIL FED. EXCISE TAX

EVANS PETITE

Especially created for the ladies — smart accessory for the handbag — the Evans Petite is designed with the same precision construction that is an Evans trade-mark.

EVANS PETITE
Golden Metal — Engine-turned
5-41-2 Retail: $3.95

EVANS PETITE
Golden Metal — Black enamel
5-2-2 blk. Retail: $3.95

EVANS PETITE
Golden Metal — Butler and Polished
5-2-2 Retail: $3.95
White Metal
5-2-3 Retail: $3.95

EVANS PETITE
Golden Metal
5-170-2 Retail: $3.95

ILLUSTRATIONS ABOUT ¾ ACTUAL SIZE

EVANS PETITE
Golden Metal—Hand-painted French Enamel (Limoges)
5-157-2 Retail: $4.95

EVANS PETITE
Golden Metal—Engine-turned design
5-421-2 Retail: $3.95

EVANS PETITE
Golden Metal—Engine-turned design
5-42-2 Retail: $3.95

PRICES SHOWN ARE FAIR TRADE RETAIL

EVANS PETITE
Golden Metal—Dark blue enamel
5-282-2 Retail: $4.95

EVANS PETITE
Golden Metal—Brown Alligator Calf
55-416-2 Alligator Retail: $3.95

EVANS PETITE
Golden Metal—Engine-turned design
5-40-2 Retail: $3.95

... AMERICA'S BEST ® Evans

PLATE 3

3

EVANS TABLE OR DESK LIGHTERS

EVANS CORNUCOPIA SMOKING SET
Table Lighter and Cigarette Holder
CC-330-22 Silver Finish Retail $17.50

EVANS VESTAL
Sterling Silver
559-18-5 Retail $25.00 (tax included)

For any room — any home — Evans table lighter designs lend beauty to any room or setting. A perfect gift and proud possession. Only Evans has the Red Seal Fuel Control which prevents fluid evaporation and provides an instant light every time.

EVANS CLASSIC
Golden Metal — Dark Blue Enamel
555-274-2 Retail $7.50
Also available in green or black

EVANS CLASSIC
Golden metal — hand painted on red French Enamel (Limoges)
555-113-2 Retail $12.50

EVANS MAGIC LAMP
550-176-22 Silver Finish Retail $9.95
550-176-2 Golden Metal Retail $9.95

EVANS CLASSIC
Golden metal — French Enamel (Limoges)
Hand-painted rose on white
555-573-2 Retail $9.95

EVANS CLASSIC
Golden metal — hand-painted on black French Enamel (Limoges)
555-204-2 Retail $12.50

EVANS CORNUCOPIA
Silver Finish
665-330-22 Retail $9.95

® Evans

PRICES SHOWN FAIR TRADE RETAIL

THERE IS NO CHARGE FOR SERVICING EVANS LIGHTERS

ILLUSTRATIONS ABOUT ⅔ ACTUAL SIZE

PLATE 4

4

ILLUSTRATIONS ABOUT ½ ACTUAL SIZE

EVANS NEW PORCELAIN LIGHTERS

Of finest porcelain—high-fired for permanence and beauty — styled to any room decor — useful and decorative. Fitted with famous Evans reliable automatic lighter — holds fuel for months.

Evans Porcelain Table Lighter
Golden Metal Top; gold and red decoration on pastel green satin glaze.
86-15-2 Retail $9.95

Evans "Rose Bowl" Table Lighter
Polished clear glass globe with removable top; weighted point for holding bud or bloom upright.
Golden Metal Top
87-1-2 Retail $7.50
(without flower)

ITEMS ON THIS PAGE FREE FROM RETAIL FED. EXCISE TAX
PRICES SHOWN FAIR TRADE RETAIL

Evans Porcelain Table Lighter
Golden metal top; gold stripes on pastel green decoration.
83-2-2 Retail $9.95

Evans Porcelain Table Lighter
Golden Metal Top; gold and red decoration on black glaze.
83-10-2 Retail $9.95

Evans Porcelain Table Lighter
Golden metal top; purple, green and gold decoration on ivory satin glaze.
86-16-2 Retail $9.95

Evans Porcelain Table Lighter
Golden metal top; Black glaze stippled with gold.
81-3-2 black Retail $7.50

Evans Porcelain Smoking Set
Table Lighter, cigarette holder and 2 ashtrays; Black glaze stippled with gold; fitted with removable Evans automatic lighter unit; gift boxed.
81A-3-2 Black Retail $12.50

® Evans

PLATE 5

5

EVANS FITTED CARRYALLS
CONTAINS LIPSTICK, VELOUR PUFF, POWDER WELL, LUCITE COMB, COIN HOLDER, SPACE FOR 10 CIGARETTES
NOTE
CARRYALL HM47-43-17 ALSO CONTAINS EVANS AUTOMATIC LIGHTER
Evans Carryall
Golden metal — Sunburst design
contains lipstick, velour puff, powder well, lucite comb, coin holder and separate space for 14 cigarettes.
H43-170-2 Retail $27.50 plus tax
Evans Carryall
Red and Green gold finish — Basketweave design
contains lipstick, velour puff, powder well, lucite comb, coin holder and separate space for 14 cigarettes.
H43-130-17 Retail $27.50 plus tax
Evans Carryall
Red and Green gold finish — Fluted design
Contains lipstick, velour puff, powder well, lucite comb, coin holder and separate space for 14 cigarettes.
H43-131-17 Retail $27.50 plus tax
Evans Carryall
White metal — Rhinestone decoration
contains lipstick, velour puff, powder well, lucite comb, coin holder and separate space for 14 cigarettes.
H43-528-3 Retail $37.50 plus tax
Evans Carryall
Red and Green gold finish — Engine turned design
Contains lipstick, velour puff, powder well, lucite comb, coin holder, Evans automatic lighter and separate space for cigarettes.
HM47-43-17 Retail $32.50 plus tax
Evans
ILLUSTRATIONS ABOUT ½ ACTUAL SIZE
PLATE 8
8

September, 1949
113
EVANS
The 20-Pak
The finest combination automatic lighter and cigarette case at any price.
Adjustable for King or Regular Cigarettes
Sweeping, pleasing lines molded with Evans expert craftsmanship — that's the new 20-PAK — an Evans fully automatic lighter and cigarette case jeweler-designed to fit the body — a combination case you'll be proud to give as a gift and to have as your own.
And what's more — the 20-PAK adjusts to hold either king or regular size cigarettes, or both — allowing an ample supply always of full, fresh smokes.
See the 20-PAK yourself — compare it with any other fully automatic lighter and cigarette case combination at any price — it's truly the finest incorporating all the exclusive design and performance features which make the new Evans automatic lighters America's Best.
AT LEADING JEWELERS AND DEPARTMENT STORES
The 20-Pak
Attractively Boxed $10 Tax Free
Gold Metal or Rhodium finish
See the 20-PAK today at leading jewelers, men's or department stores — it's the finest — an Evans.
The 20-PAK fits your pocket without bulging — yet carries a day's supply for the average smoker, king or regular size — always on hand — always just like new.
... AMERICA'S BEST ® Evans is elegance
Evans Case Company, North Attleboro, Mass.
New York City • Chicago • Los Angeles • Dallas • Cleveland • Boston
CREATORS OF FAMOUS EVANS FITTED HANDBAGS, AUTOMATIC LIGHTERS AND POWDER BOXES

EVANS FITTED HANDBAGS
Smartly tailored fashions crafted of the finest materials — clasped by Evans frames and fitted with matched accessories; each in its own snug pocket, always handy for instant use . . . powder pox with Evans exclusive leak-proof sifter, soft velour puff . . . one piece pop-up lipstick holder . . . cigarette case and sensational new Evans automatic lighter.
Black, Brown or Navy calf with matching satin lining and change purse. Golden metal finished frame and fittings; deep-cut engine-turned design. Fitted with cigarette case, Evans automatic lighter, powder box, lipstick holder and comb. Size 9 x 6 inches.
1868 Retail $41.50 plus tax
Black, Brown or Navy calf with matching satin lining and change purse. Golden metal finished frame and fittings; deep-cut engine turned design. Fitted with powder box, lipstick holder and comb. Size 7 x 5 inches.
X1992 Retail $47.50 plus tax
Black, Brown or Navy calf with matching satin lining and change purse. Golden metal finished frame and fittings; deep cut engine-turned design. Fitted with cigarette case, Evans automatic lighter, powder box, lipstick holder and comb. Size 8 x 7 inches.
1999 Retail $56.00 plus tax
ELEGANCE EVANS
® Evans is elegance
PLATE 9
9

AUTOMATIC Evans LIGHTERS
SENSATIONALLY PRICED
Formerly $4.00 & $4.50
NOW
$2.95 TAX INC.
Unconditionally Guaranteed
RED SEAL FUEL CONTROL
Evans is elegance
GIVES 1000's OF EXTRA LIGHTS
ASSORTMENT NO. CD2
Retailers' Cost $21.60 Retail $35.40
New value-packed easel back window or counter display. 14 x 18 inches with one dozen Evans automatic lighters to retail at $2.95 each. These assorted chrome and golden metal finish lighters are covered with a durable, washable, simulated leather in ostrich, alligator, and snake-skin patterns as shown.
® Evans

SEE DETAILS ON EVANS EXCLUSIVE RED SEAL AND OTHER FEATURES ON REVERSE SIDE
HOME LINES
LIGHT UP YOUR TRADE-MARK with a NEW EVANS
Delight friends and influence customers - garner new business with your own trade-marked (in colors) new Evans fully automatic lighters or powder boxes.
Here's a sure fire colorful way to gain constant name and product impression - and it's easy as A. B. C. Simply send inscription and quantity desired - at no cost to you we'll submit design and price (quantity minimum 250).
Let America's Best Nationally Advertised Evans fully automatic lighters spark your sales and good will daily. Here's a gift that will always be remembered, used with thankful satisfaction.
The Exclusive Evans Red Seal prevents fluid evaporation and overflow - one of the many features which make Evans lighters the best gift.
READ ABOUT EVANS EXCLUSIVE RED SEAL AND OTHER FEATURES ON REVERSE SIDE

Carryalls & Purses
Shirley Temple
THE STORY OF SEABISCUIT
A Warner Bros. Production
Finishing Touch to Evening Glamour...
EVANS bracelet-handled Coronation Carryall — fitted with after-five indispensables. Brightest rhodium reflects a crown of rhinestones, heralds the elegance within.
EVANS CASE CO.
No. Attleboro, Mass.
Sales Offices:
Inside, jewelet-designed automatic "pop-up" lipstick, covered powder well, puff, comb and coin holder plus separate space for cigarettes . . . (hanky or keys if you prefer).
At leading jewelry and department stores.
Evans is elegance
EVANS
Coronation Carryall — $36.50*
Other carryalls from $25.*
CREATORS OF FAMOUS EVANS FITTED HANDBAGS, AUTOMATIC LIGHTERS, CIGARETTE CASES AND POWDER BOXES
1949
139

THE HASTY HE
EVANS
THE COMMUTER
Evans is elegance
EVANS CASE CO., No. Attleboro.

Evans "Lamp-Liter" For his office or den. A clever and useful gift ... a fine lamp with 3-way socket and an EVANS Bone China Table Lighter in Onyx White, hand-decorated with game birds. Actual size of lamp - 15 inches high. Entire base of lighter is fuel reservoir holding a year's supply of lights.
LP815-74-2 White Retail $25.00
Also available with plain lamp shades in Maroon, Green or Beige (specify color).
Evans

A
B
D
E
Can classic elegance be inexpensive?
Evans® says, of course . . . with one simple enduring style offered in these elegant variations: (A) Brushed chrome. (B) Gun metal and gold, genuine lizard case. (C) Gun metal and gold finish. (D) Hand textured swirl, gold finish. (E) Personalized hand textured gold finish. Look for Evans lighters wherever smokers' supplies are sold. American made . . . lifetime repair of mechanism fully guaranteed. Priced from $4.95 to $12.95.
NEW EVANS EYELITER®
(Don't fuel till you see the white of its eye)
The newest, most unique pocket lighter made. Red eye turns white to tell you when to refill. Windproof, adjustable flame, lifetime guarantee. Available in black or chrome finish. Individually packaged for gift giving. Only $9.95.
Evans® BUTANE LIGHTERS
Evans Case Company, Plainville, Massachusetts 02762
14.8.5
This advertisement appears in Look and Life Magazines December, 1968

Evans new
eyeliter
AUTOMATIC BUTANE LIGHTER
3-Pak Display
BUTANE LIGHTER & FUEL ASSORTMENT
eyeliter®
RED EYE TURNS WHITE
Tells you when to refuel
AUTOMATIC BUTANE
ADJUSTABLE - WINDPROOF
LIFETIME GUARANTEE
#56/954
2—56/500/01 $9.95 Chrome Finish Lighters $19.90
1—56/500/05 $9.95 Black Enamel Finish Lighter 9.95
3—$1.00 Cans Fits-All Fuel 3.00
TOTAL RETAIL VALUE $32.85
DEALER COST 19.71
YOUR PROFIT $13.14
Unique pilferproof display features the amazing new automatic Evans Eyeliter with Red Eye that fades as fuel decreases and turns white when fuel in lighter is exhausted.
Display is only 15" high and fits neatly on counter top or window.
Shows and Sells 3 $9.95 Butane Lighters and 3 $1.00 Cans Fits-All Butane Fuel.
EVANS CASE CO. PLAINVILLE, MASS.
All resale prices are suggeste
Printed in U.S.

Chapter 4

Pocket Lighters

Petite pocket lighter in mother-of-pearl with small colored stones. Circa 1950. $25-40.

Trig-a-Lite model pocket lighter, chrome-plated with black enamel panels. Circa 1935. $25-150.

Trig-a-Lite pocket lighter in chrome-plate with hobnail pattern. $15-125.

Banner pocket lighter, gold-plated with mother-of pearl panels. Circa 1950. $25-75.

Petite pocket lighter in gold-plate with multicolored stones in a floral pattern. $15-35.

Gold-plated pocket lighter with red snakeskin covering on Evans pocket lighter display stand. The display stand is rare. Lighter: $15-40. Stand: $50.

Chrome-plated Banner pocket lighter with windscreen. Circa 1950. $15-40.

Left: Chrome-plated Banner with brown enamel side panel. $15-40. Right: Gold-plated Banner in white enamel with a floral pattern. $25-75.

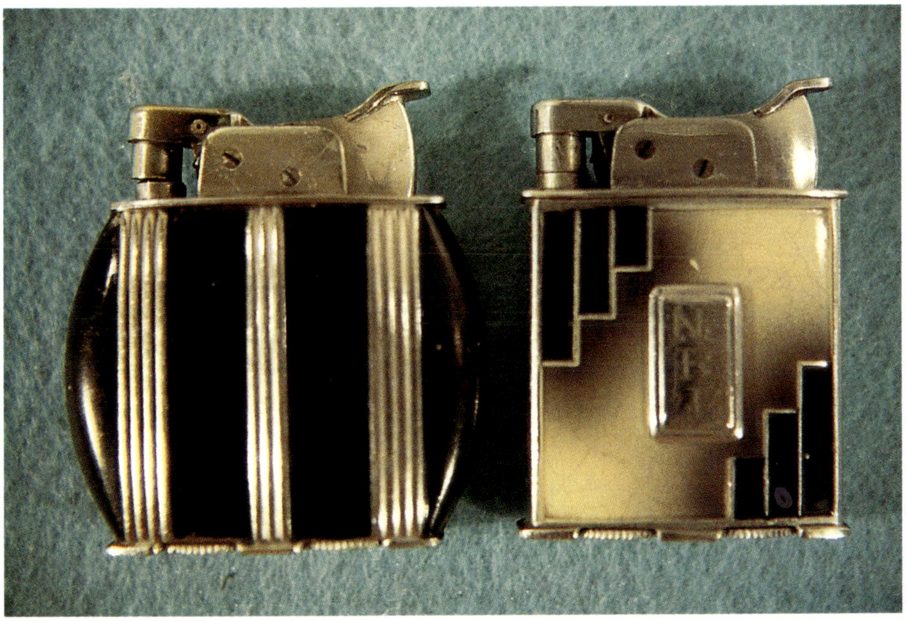

Two enameled Trig-a-lite pocket lighters in chrome plate. Circa 1930. Left: $20-150. Right: $20-225.

Cloisonné-enameled lift-arm pocket lighter in black and white enamel with a floral pattern. Enameled lift-arms are rare and much sought after by Evans collectors. Circa 1928. $350-700.

Left: Banner pocket lighter in gold-plate and mother-of-pearl. $25-75. Right: Pocket lighter in gold-plate with hobnail pattern and large screws. Evans pocket lighters with large screws are unnamed but are sometimes referred to as the "Spitfire" by Evans collectors. This lighter was manufactured after the Trig-a-lite model and before the Banner model. This model was discontinued in the mid-1940s. $15-40.

The Supreme model pocket lighter with a kneeling nude in multicolored enamel. $50-100.

Banner pocket lighter, "International Harvester Safety Award." $15-40.

Banner pocket lighter in gold-plate advertising Cott Beverages. Evans advertising pocket lighters are uncommon. Models in mint condition are much sought after by Evans collectors. $50-100.

Banner pocket lighter with a Cadillac logo. $40-75.

Banner pocket lighter in chrome-plate with alligator-grain calfskin covering. $15-40.

Golden Prince model pocket lighter with a red gold finish. Circa 1950. $15-40.

Banner pocket lighter in gold-plate with black calfskin covering. $15-40.

Supreme pocket lighter in white cloisonné enamel with a floral pattern. $25-100.

Gold-plated Banner pocket lighter in green enamel advertising Rohr Aircraft. $40-100.

Trig-a-lite pocket lighter in chrome-plate with a green enamel art deco pattern. $15-225.

Trig-a-lite pocket lighter with chrome and green enamel design. $40-200.

Trig-a-lite pocket lighter in black enamel with an art deco design. $15-225.

Petite pocket lighter with white calfskin covering in original box. $15-40.

Banner pocket lighter in chrome-plate. $15-40.

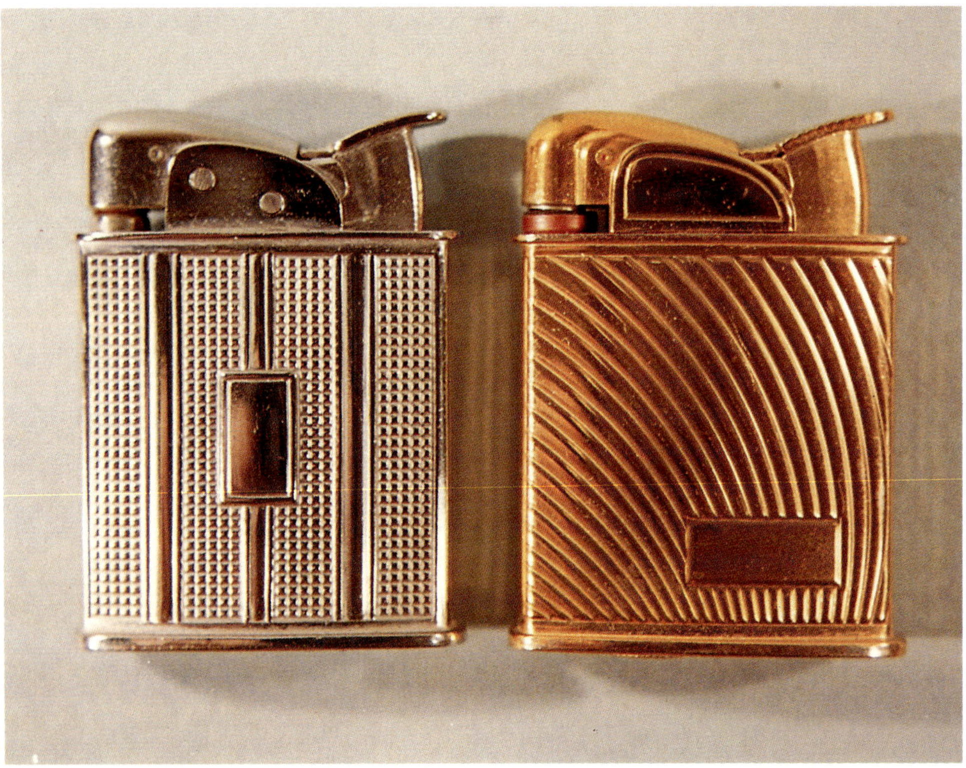

Two Banner pocket lighters, one in chrome-plate and one in gold-plate. $15-40 each.

Pocket lighter advertising Camel Cigarettes. $25-175.

Roller-bearing model pocket lighter in green enamel with a trio of puppies. The Roller-bearing was the predecessor of the Trig-a-lite. Circa 1928. $15-225.

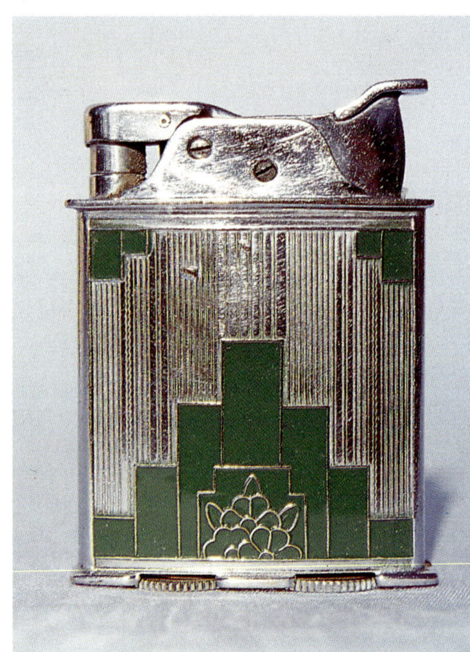

Roller-bearing pocket lighter in a chrome and green enamel art deco pattern. $15-225.

Butane pocket lighter with gun-metal finish in a lizard skin pouch. This lighter could have been manufactured by Evans Case or the Hilsinger Company. $15-35.

Petite pocket lighter in gold-plate with mother-of-pearl panels. $20-40.

Warwick pocket lighter in gold-plate and black enamel design. This is a Banner pocket lighter with a design name of "Warwick." $15-40.

Trig-a-lite pocket lighter with a chrome and black enamel design. $15-150.

Petite pocket lighter in cloisonné enamel with a floral pattern. $15-40.

Banner pocket lighter in gold-plate with original box and price tag. $15-40.

Enameled lift-arm with oriental design in green and black. $350-700.

Butane pocket lighter marked "EVANS" on the bottom. This lighter was made for Hilsinger by the makers of the "Prince" pocket lighter, which is identical in design. Circa 1965. $20-40.

Trig-a-lite pocket lighter in green enamel with an art deco design. This particular design came in a variety of colors and was also sold with a matching cigarette case. $25-225.

Cloisonné-enameled lift-arm in blue and black regimental stripes. $350-700.

Gold-plated Banner pocket lighter with engraved design. $15-40.

Two calfskin covered Banner pocket lighters. Circa 1950. $15-40.

Left: Gold-plated Banner with a mother-of-pearl panel. $15-40. Right: Chrome-plated Banner with initial plate. $15-40.

Banner pocket lighter in gold-plate with a black and white snakeskin covering. $15-40.

Large pocket lighter with a black crackle finish. I believe that this lighter was made for use by the military during World War II. It is easy to find, which is an indicator of a large production quantity. $10-25.

Roller-bearing pocket lighter with a shark-grain effect in a modernistic design. $15-225.

Petite pocket lighter in gold-plate with multicolored stones. $15-40.

Supreme pocket lighter in rhodium-plate with a jeweled crown decoration. $25-75.

Supreme pocket lighter with a nude design in enamel. $50-100.

Clipper pocket lighter in a black enamel art deco design. Circa 1940. $25-175.

Banner pocket lighter in blue and white enamel advertising Harrah's Club. $40-100.

Kingston pocket lighter with original price tag setting on an Evans pocket lighter stand. $15-40 for the lighter.

Trig-a-lite pocket lighter in black enamel with an oriental decoration in chrome plate. The lighter pictured is in mint condition. $15-200.

Chrome-plated Banner pocket lighter with a comical theme. $25-60.

Petite pocket lighter in white cloisonné enamel with a floral design. $25-40.

Two gold-plated Banner pocket lighters. The lighter on the left is dark blue enamel with gold diagonal stripes. The lighter on the right has a brown calfskin covering and an initial plate. $15-40 each.

Gold-plated Petite pocket lighter with a brown snakeskin covering. This type of Petite lighter (lighter mechanism covered) was sold with a matching cigarette case. $15-40.

Chrome-plated enameled Banner pocket lighter advertising Glen Gery Brick. $40-100.

Chrome-plated Clipper pocket lighter with advertising for Lucky Strike cigarettes. $40-200.

Gold-plated Kingston pocket lighter with a cameo decoration in original box. $20-40.

Trig-a-lite pocket lighter in chrome and black enamel with an art deco pattern. $25-225.

Gold-plated Banner pocket lighter with yellow cloisonné enamel in a floral design. $25-100.

Chrome-plated Banner pocket lighter with windscreen. $15-35.

Chrome-plated Banner pocket lighter advertising Bourbon de Luxe Whiskey. $50-100.

Gold-plated Petite pocket lighter in sunburst design. $15-30.

Left:
Cloisonné-enameled lift-arm in rose and white enamel with a floral pattern. This lighter has the identical pattern on the reverse which makes it very rare. $400-$1,000.

"Warwick" design pocket lighter in chrome-plate and black enamel. $15-40.

Below:
Left: Gold-plated Supreme pocket lighter in white enamel with a floral pattern. $50-125. Right: Gold-plated Supreme pocket lighter in black enamel. $15-40.

Enameled Banner pocket lighter in Trojan Horse design with matching cuff links. $50-100 set.

Gold-plated Petite pocket lighter with acrylic panels in a quartz and gold design. $20-40.

Below:
A group of Supreme pocket lighters, all with nudes. $50-100 each.

An uncommon, tall Trig-a-lite pocket lighter in chrome-plate and black enamel. Rare in mint condition. $25-150.

Chrome-plated pocket lighter in white enamel with the description, "General." $15-50.

Gold-plated Banner pocket lighter with the "UNION 76" logo. $50-100.

Gold-plated Banner pocket lighter in black, white, and gold enamel. $15-50.

Chrome-plated Banner pocket lighter in a pattern that was also used on case-lighters. The lighter pictured is in mint condition with its original box. $15-40.

Trig-a-lite pocket lighter with a brown enamel and chrome-plate art deco design. $20-225.

Trig-a-lite pocket lighter in a green enamel and chrome-plate art deco pattern. $25-225.

Chrome-plated Supreme pocket lighter with a military logo in green, gold, and white enamel. $15-50.

Gold-plated supreme pocket lighter advertising radio station WSID. Evans made pocket lighters for many radio stations across the country. $15-40.

Chrome-plated Supreme pocket lighter advertising Lucky Strike cigarettes. $50-100.

Left:
Chrome-plated Trig-a-lite pocket lighter with a beautiful smoky green enamel pattern. $20-200.

Banner pocket lighter in sterling silver with an engraved design. $75-150.

Chrome-plated pocket lighter in an art deco pattern with large screws and a windscreen. $15-40.

Gold-plated Banner pocket lighter with a man serenading two women. This same pattern appears on case-lighters and perfumers. $25-100.

Two Flip-Top model pocket lighters, one in red and black enamel, the other with a black calfskin covering. Circa 1955. $40-75 each.

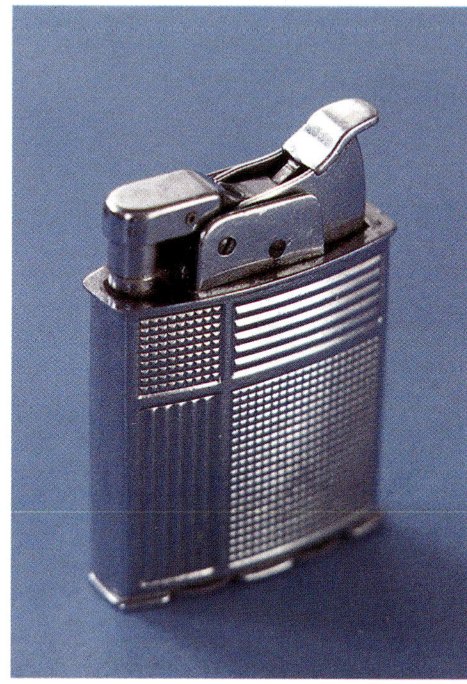

Chrome-plated Trig-a-lite pocket lighter with a hobnail/art deco pattern. $25-100.

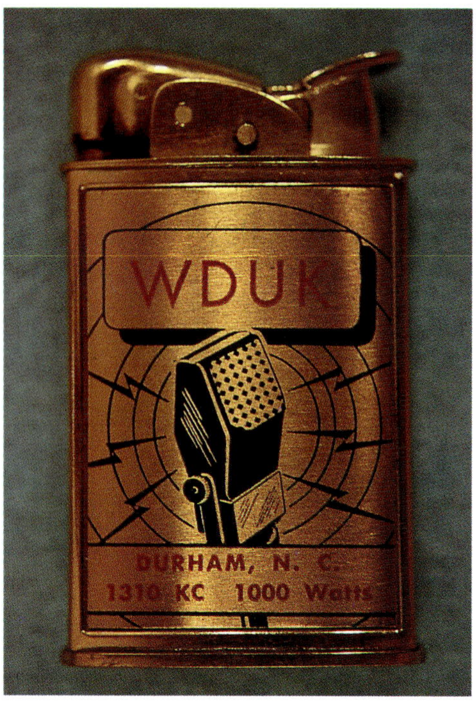

Gold-plated Supreme pocket lighter advertising radio station WDUK. $15-40.

Left: Banner pocket lighter in gold-plate and dark blue enamel with a fleur-de-lis pattern. $15-40. Right: Gold-plated Banner pocket lighter with a Trojan warrior's helmet and sword. $15-40.

Chrome-plated Supreme pocket lighter with a ribbed design. $15-35.

Enameled lift-arm in chrome-plate with a checkerboard pattern. $300-600.

Gold-plated Banner pocket lighter with a ribbed pattern in red enamel. $15-40.

Chrome-plated pocket lighter in white, red, and black enamel advertising Pepperell Fabrics. $50-100.

Two gold-plated Banner pocket lighters with calfskin coverings. $15-40.

Gold-plated Kingston model pocket lighter in a basket weave pattern with original box. $20-40.

Cloisonné-enameled lift-arm in sky blue and white with a floral pattern. $350-700.

Chrome-plated petite Trig-a-lite with a hobnail/art deco pattern. $15-125.

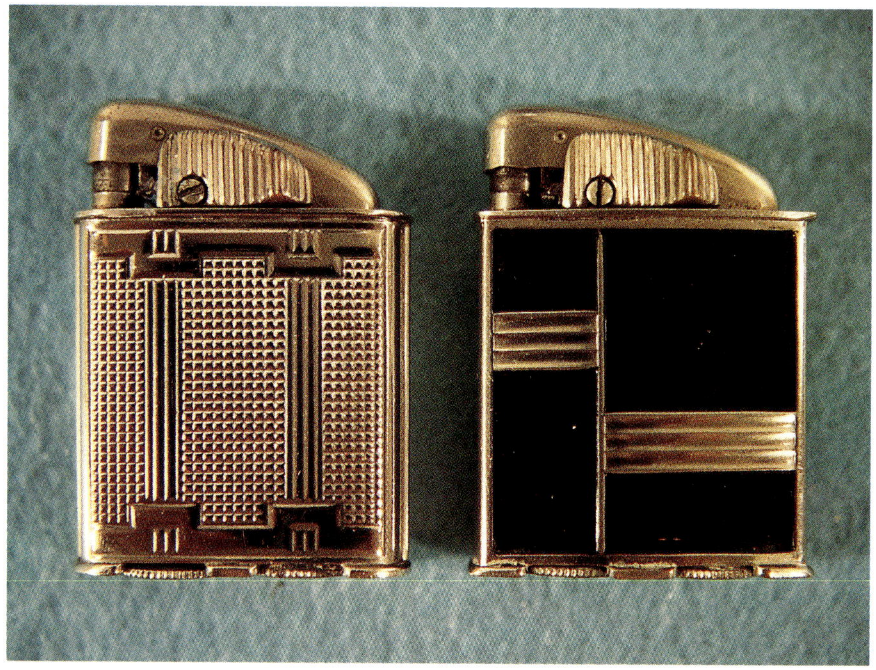

Left: Chrome-plated Clipper model pocket lighter with a hobnail/art deco pattern. $25-100. Right: Clipper pocket lighter with a black enamel and chrome-plate art deco pattern. $25-200.

Supreme pocket lighter in chrome-plate with a light blue enamel panel. $15-40.

Gold-plated Petite pocket lighter with light green acrylic panel. $15-40.

Chrome-plated Banner pocket lighter in white enamel with the words "Crossed International Date Line, 1952." $25-75.

Evans' first model pocket lighter called the "Automatic." This lighter is seldom seen in this rare enameled pattern. $300-700.

Banner pocket lighter in mother-of-pearl and oyster shell. $25-75.

Supreme pocket lighter in gold-plate with a ribbed design. $15-50.

Cloisonné-enameled lift-arm in black and white with a floral pattern. $350-700.

Trig-a-lite pocket lighter in black enamel with a gazelle design. $25-225.

Trig-a-lite pocket lighter in chrome-plate and tortoise-shell enamel. $15-150.

Small Trig-a-lite pocket lighter in chrome-plate with a hobnail pattern. $15-100.

Banner pocket lighter in gold-plate and white enamel with a bird and leaf design. $15-50.

Supreme pocket lighter in gold-plate advertising radio station WHEB. $15-35.

Roller-bearing pocket lighter in chrome-plate with a two-tone blue enamel art deco design. $25-200.

Sterling silver pocket lighter with a hobnail pattern. $50-150.

Three leather-covered pocket lighters. $10-30 each.

Trig-a-lite pocket lighter in chrome-plate and enamel showing a man serenading two women. $75-350.

Roller-bearing pocket lighter in chrome-plate with an engine-turned design. $15-100.

Petite pocket lighter in gold-plate and mother-of-pearl. $15-35.

Trig-a-lite in chrome-plate advertising Lucky Strike cigarettes. $50-200.

Supreme pocket lighter in chrome-plate and enamel showing a fishing fly. $15-60.

Petite pocket lighter in gold-plate with a ribbed design. $15-30.

Gold-plated Banner pocket lighter with a mother-of-pearl panel. $25-75.

Pink leather-wrapped Petite with a large, light blue stone. $15-30.

Left: Gold-plated Banner in herringbone pattern. $15-40. Right: Gold-plated Banner with mother-of-pearl panels. $25-75.

Chrome-plated Breeze King model pocket lighter. Circa 1955. $20-60.

Left: Gold-plated Banner showing the state of Alaska. $15-40. Right: Gold-plated Banner with a panel resembling sandstone. $15-40.

Trig-a-lite pocket lighter in chrome-plate and black enamel with an artistic design. $25-200.

Red gold-plated pocket lighter with a diamond pattern. $15-40.

Cloisonné-enameled lift-arm in an unusual pattern. $350-700.

Cloisonné-enameled lift-arm with blue and white panels. $350-700.

Gold-plated Banner pocket lighter advertising California Cornice, Steel, & Supply Corp. $50-100.

Supreme pocket lighter in chrome-plate and enamel advertising the Thrift Novelty Company. $50-100.

Ladies' Flip-Top pocket lighter in white cloisonné enamel with a floral design. An uncommon lighter. Circa 1955. $100-200.

Trig-a-lite pocket lighter in chrome-plate and brown enamel. $15-150.

Cloisonné-enameled lift-arm with green and white panels. $350-700.

Breeze King in gold-plate with leather-wrap. $40-75.

Gold-plated Banner showing an Olds Runabout, 1903. $15-40.

Banner pocket lighter advertising Ebony suntan lotion. $50-100.

Left: Supreme pocket lighter with a leaping trout. $15-40. Right: Supreme pocket lighter with the Cadillac logo in enamel. $15-75.

Gold-plated Petite pocket lighter with mother-of pearl and tortoise-shell panels. $15-30.

Gold-plated pocket lighter made by Evans for Hickock, a leading jewelry manufacturer in the 1950s. The lighter is marked "Hickock" on the bottom. Rarely seen. $40-75.

Cloisonné-enameled Clipper pocket lighter with a floral pattern. $150-300.

Gold-plated Banner pocket lighter with an enameled floral design. $25-50.

Two Banner pocket lighters wrapped in calfskin. $15-40.

Cloisonné-enameled Roller-bearing pocket lighter with an art deco design. This same design appears on an Evans perfumer. $100-300.

Gold-plated Banner pocket lighter with bowling ball and pins. $15-40.

Roller-bearing pocket lighter with black and red enamel panels and shark-grain effect. $25-225.

Cloisonné-enameled lift-arm with black and red regimental stripes. $350-700.

Cloisonné-enameled lift-arm with blue and white panels and a floral design. $350-700.

Chrome-plated pocket lighter with a basket weave design. $15-35.

Early automatic pocket lighter in chrome-plate with a hammered finish. $50-150.

Gold-plated Petite pocket lighter with a jeweled design and keychain. $25-50.

Chrome-plated lift-arm showing a man in top hat and tails. $75-200.

Gold-plated Banner pocket lighter showing Ford's First Car, 1896. $15-40.

Gold-plated Banner pocket lighter wrapped in calfskin, with a fleur-de-lis pattern. $15-40.

Supreme pocket lighter in chrome-plate advertising Camel cigarettes. $50-100.

Cloisonné-enameled lift-arm in blue and white with a floral design. $350-700.

Petite pocket lighter, leather-wrapped with a floral pattern made of seashells. $15-30.

Petite pocket lighter in gold-plate, wrapped in snakeskin. $15-30.

Supreme pocket lighter in gold-plate advertising Beneke Tuffy. $50-100.

Banner pocket lighter in sterling silver. $75-150.

Gold-plated pocket lighter with a basket weave pattern. $15-40.

Small Trig-a-lite pocket lighter in chrome-plate and white enamel. This lighter is in mint condition. $20-150.

Chrome-plated lift-arm wrapped in animal skin. $75-150.

Roller-bearing pocket lighter with elephants in enamel. A beautiful lighter that is very rare in this pattern. $300-700.

Gold-plated Petite pocket lighter wrapped in animal skin. $15-30.

Gold-plated Petite pocket lighter with rhinestones. $25-50.

Two leather-wrapped Petite pocket lighters. $15-30 each.

Cloisonné-enameled lift-arm in green and white regimental stripes with a floral pattern. $350-700.

Red gold-plated Supreme pocket lighter with a ribbed design. $15-40.

Cloisonné-enameled lift-arm with black and white panels. $350-700.

Gold-plated Supreme pocket lighter with a ribbed design. $15-40.

Two gold-plated Banner pocket lighters. $15-40 each.

Two enameled pocket lighters with large screws. $15-75.

Cloisonné-enameled lift-arm with blue and white panels. $350-700.

Gold-plated Banner pocket lighter advertising Manila Rum. $50-100.

Gold-plated Petite pocket lighter, leather-wrapped, with a butterfly pattern in imitation pearls. $15-30.

Cloisonné-enameled lift-arm with green and white squares and a floral pattern. $350-700.

Gold-plated Supreme pocket lighter advertising Old Gold cigarettes. $50-100.

Two enameled Banner pocket lighters, one with "FOE" logo and the other with "CLAROSTAT" logo. $50-100 each.

Roller-bearing pocket lighter in enamel. Very rare in this pattern. $300-700.

Roller-bearing pocket lighter in chrome-plate and red enamel with an art deco design. $25-225.

Chrome-plated Trig-a-lite with an engine-turned design. $15-150.

Petite pocket lighter in white cloisonné enamel with a floral pattern. $25-50.

Gold-plated, leather-wrapped "Pony" lift-arm. Presence of the red dot usually is an indicator that the lighter is in mint condition, as the dot gradually wears off with use of the lighter. $40-150.

Nickel-plated "Pony" lift-arm with a hammered finish. $60-100.

Chrome-plated pocket lighter in a hobnail pattern with large screws. $15-40.

Chrome-plated Roller-bearing pocket lighter with an art deco design in blue enamel. This lighter is in mint condition. $25-300.

Chrome-plated Trig-a-lite pocket lighter with a hobnail/art deco pattern. $15-100.

Cloisonné-enameled lift-arm in a rainbow pattern which is very rare. $500-1,000.

Left: Chrome-plated Banner pocket lighter with a calfskin panel. Right: Chrome-plated pocket lighter with an engine-turned design. $15-40 each.

Flip-Top pocket lighter in gold-plate with tortoise-shell finish. Circa 1955. $40-75.

Gold-plated Petite pocket lighter wrapped in pink calfskin with small rhine-stones. $15-30.

Early Flip-Top pocket lighter in gold with a basket weave pattern. This lighter was given by Bing Crosby to his wife Dixie. The lighter is inscribed "Dixie, Xmas, 1942." *Collection of Guy Nishida.*

A group picture of the Evans enameled pocket lighters from the collection of PLPG President Ron Eyerkuss. Ron has the finest known collection of Evans enameled pocket lighters.

Chapter 5
COMBINATION SETS

Combination set consisting of a Breeze King pocket lighter, tie bar, and cuff links, all in black enamel. $75-150 set.

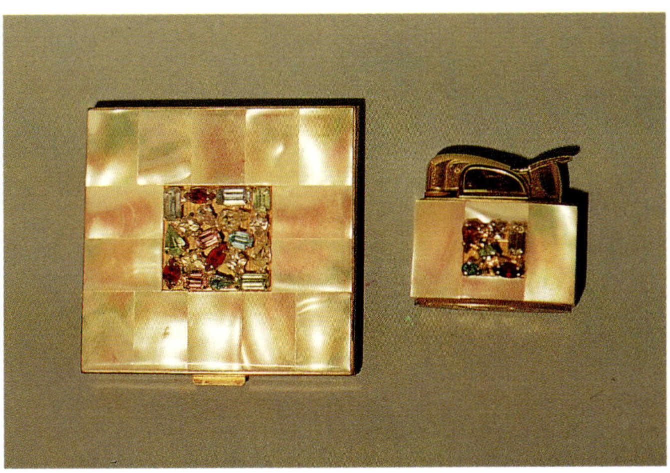

Petite pocket lighter and matching cigarette case in mother-of-pearl with multicolored stones. $75-150 set.

Petite pocket lighter and matching cigarette case in mother-of-pearl and tortoise-shell. $75-150 set.

Four-piece matching set in gold-plate with a herringbone pattern.
$75-150 set.

Gold-plated cigarette case with built-in cigarette lighter (open position).
$25-75.

The same cigarette case in closed position.

Trig-a-lite pocket lighter with matching cigarette case, both in chrome-plate and enamel. $75-250 set.

Gold-plated Banner pocket lighter in ribbed design with matching cigarette case. $30-75 set.

Cloisonné-enameled lift-arm in white and green with matching Carryall. $700-1200 set.

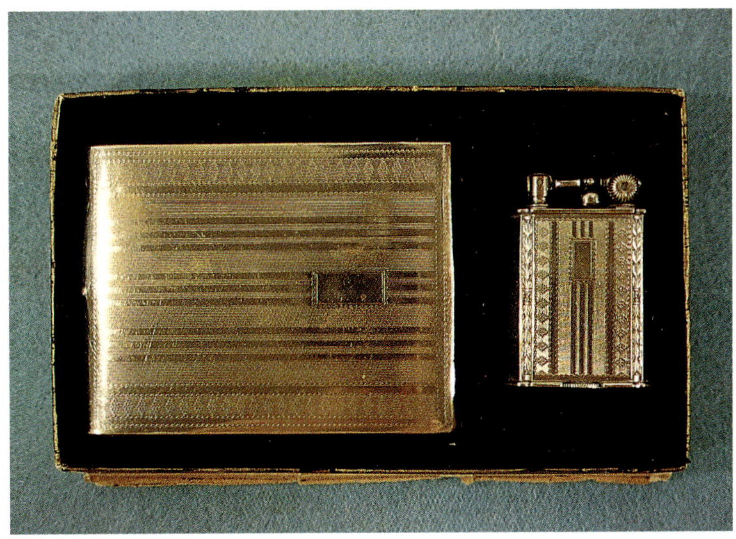

Chrome-plated, engine-turned lift-arm with matching cigarette case. $200-400.

Petite pocket lighter in white cloisonné enamel with matching cigarette case. $75-150 set.

Gold-plated Roller-bearing pocket lighter in cloisonné enamel with a matching compact. $250-450 set.

Gold-plated pocket lighter with matching cigarette case, both in a floral pattern. $50-100 set.

Gold-plated Petite pocket lighter with matching cigarette case, both with blue stones. $75-150 set.

Gold-plated Banner pocket lighter in yellow enamel with a matching bracelet. $75-150 set.

Two-tone gold-plated pocket lighter in a basket weave pattern with matching cigarette case. $35-75 set.

Four-piece matching set in brown enamel and brown calfskin. $50-125 set.

Trig-a-lite pocket lighter in black and gray enamel with matching cigarette case. $100-250 set.

Three-piece matching set in white cloisonné enamel in a floral pattern. $200-300 set.

Roller-bearing pocket lighter in a green enamel art deco design with matching cigarette case. $250-600 set

Gold-plated Banner pocket lighter in herringbone pattern with matching cigarette case. $40-75 set

Three-piece matching set consisting of a cigarette lighter, tie bar, and cuff links made by Evans for Hickock. $60-100 set.

A gorgeous set consisting of a Roller-bearing pocket lighter with matching cigarette case and cigarette holder, all in white cloisonné enamel. Extremely rare. $1,000-2,000 set.

Petite pocket lighter in mother-of-pearl and imitation pearl decorations with matching cigarette case. $75-150 set.

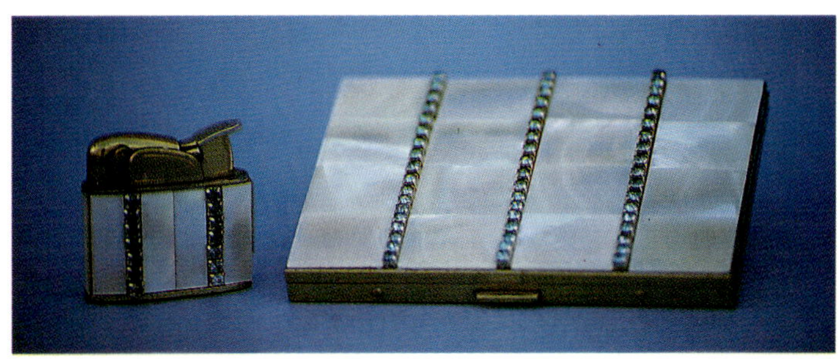

Petite pocket lighter in mother-of-pearl with light blue stones and matching cigarette case. $75-150 set.

Petite pocket lighter, gold-plated, with matching cigarette case. $40-75 set.

Banner pocket lighter in black enamel with a crown pattern and matching cigarette case. $100-250 set.

Petite pocket lighter in light blue and white enamel with matching cigarette case. This same pattern was used on Evans compacts. $75-150 set.

Petite pocket lighter in gold-plate with matching cigarette case that has a built-in watch. Rare. $200-300 set.

Chapter 6

Cigarette Case-Lighters

The Boardman-Polando Commemorative Lighter

This Evans cigarette case-lighter was a souvenir of the welcome-home celebration held in Hyannis, Massachusetts, for two fliers after their world-record-breaking flight from Floyd Bennett Field, New York, to Istanbul, Turkey, June 28-30, 1931. This was a remarkable flight which seems to have been eclipsed in most aviation history books by the round-the-world flight of Pangborn and Herndon which began the same day from the same airfield, Floyd Bennett.

Boardman-Polando commemorative lighter. $100-200.

Russell Boardman was the pilot, John Polando was his co-pilot/mechanic. John used to say, "Russell asked me how much I weighed. When I told him '120 pounds,' he said, 'O.K., you'll do.'" As gasoline weighs six pounds per gallon, John's light weight meant that the plane could carry another five or ten gallons, which, as it turned out, were needed after some fifty hours in the air. Very little fuel remained when they landed.

The plane, which was Boardman's, was a single-engine Bellanca Special called the "Cape Cod." It was a flying gas tank. On takeoff, it carried more than its own weight in fuel. John Polando's seat was a gas tank. The flight itself was very difficult, as the weather was awful with fog across the Atlantic and most of Europe. They circled the Alps waiting for clearance and eventually landed after more than two days aloft, an amazing feat considering how primitive their equipment was.

There were many celebrations after their flight, but one of the most heartfelt was the one for the "Cape Cod" and its crew of local heroes when they flew into Hyannis and for which the lighter/case was a memento.

Boardman lost his life shortly after his record-breaking flight in the crash of his GeeBee racer during a Cleveland air race.

John Polando helped a great many young people to become active in aviation. In 1984, he helped to set up an annual scholarship to help high-school students who were planning careers in aviation. As of 1994, this scholarship had awarded a total of $13,000 to more than a dozen young men and women.

Roller-bearing case-lighter in lavender and black enamel with a floral design. A rare and beautiful lighter. $100-600.

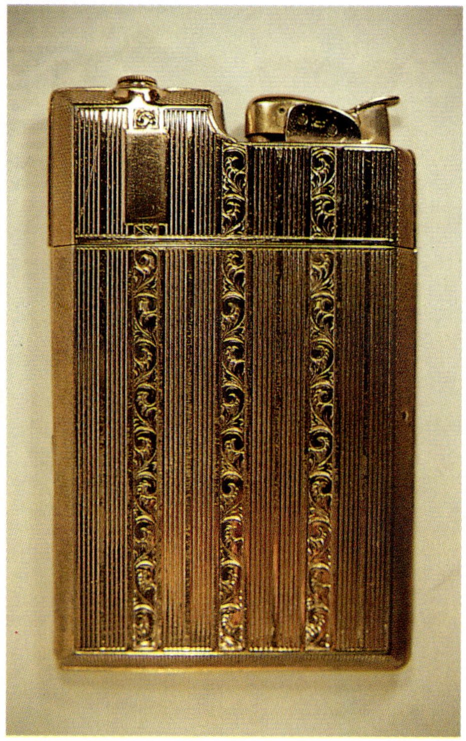

Gold-plated 20-Pak model case-lighter with engine-turned design. $25-75.

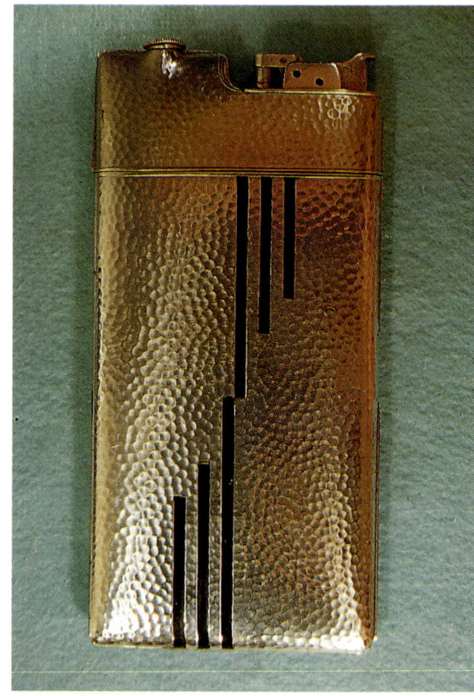

Tall chrome-plated case-lighter with black enamel stripes and a hammered finish. $25-50.

Trig-a-lite case-lighter in black and lizardskin enamel. $50-150.

Chrome-plated 20-Pak case-lighter with ribbed design. $25-50.

Chrome-plated case-lighter with windscreen in a herringbone, hobnail, and black enamel design. $25-125.

Clipper case-lighter in a herringbone, hobnail, and black enamel design. $50-175.

Black and white enamel case-lighter with built-in compact. $40-250.

Chrome-plated Monterey model case-lighter with a sunray design. $25-50.

Gold-plated case-lighter with a nude design. $100-250.

Trig-a-lite case-lighter in a deep purple enamel. $25-150.

Tall case-lighter in black and lizardskin enamel with built-in compact. $50-225.

Chrome-plated case-lighter with a hobnail/art deco pattern. $25-100.

Chrome-plated case-lighter with a hammered finish. $20-75.

Blue cloisonné-enameled case lighter with a built-in compact. $35-250.

Trig-a-lite case-lighter with a woodblock enamel design. $25-175.

Tall yellow-enameled case-lighter with a built-in compact. $100-300.

Chrome-plated Trig-a-lite case-lighter with a hammered finish. $40-125.

Chrome-plated case-lighter with a military aviation emblem. This same case pattern also appears on an Evans pocket lighter. $20-75.

Chrome-plated Trig-a-lite case-lighter with black enamel stripes and a hammered finish. $25-100.

Black enamel case-lighter with a compact lid in white cloisonné enamel with a floral design.

Chrome-plated case-lighter with windscreen and a hobnail/art deco pattern. $20-75.

Gold-plated Monterey case-lighter with a built-in watch. $100-300.

Tall gold-plated case-lighter with an enameled design near the bottom. A beautiful lighter. $100-300.

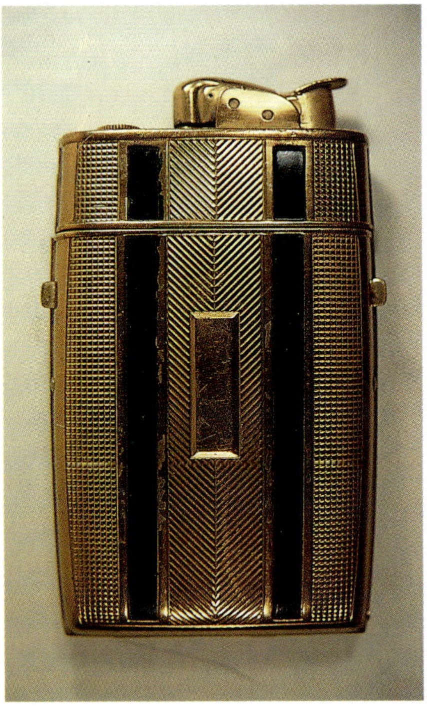

Gold-plated case-lighter with a herring-bone pattern and black enamel stripes. $35-75.

Trig-a-lite case-lighter in red and maroon enamel in an art deco pattern. $25-225.

Maroon enameled case-lighter with built-in compact. $25-225.

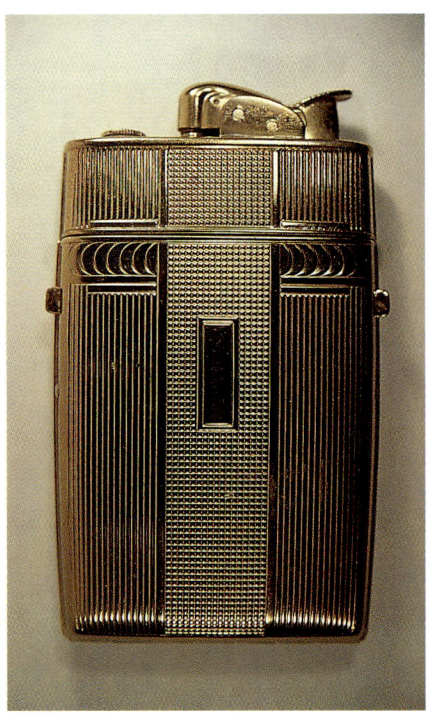

Gold-plated case-lighter with a hobnail/art deco pattern. $25-100.

Two-tone gold-plated case-lighter with two leaping gazelles. $40-150.

Chrome-plated, black enameled case-lighter with a design known as "Warwick." $20-50.

Black and white-enameled case-lighter with a built-in compact. $25-225.

Case-lighter in chrome-plate with an enameled burled maple finish. $25-175.

Tortise-shell-enameled case-lighter in gold-plate with a Clipper mechanism. $50-225.

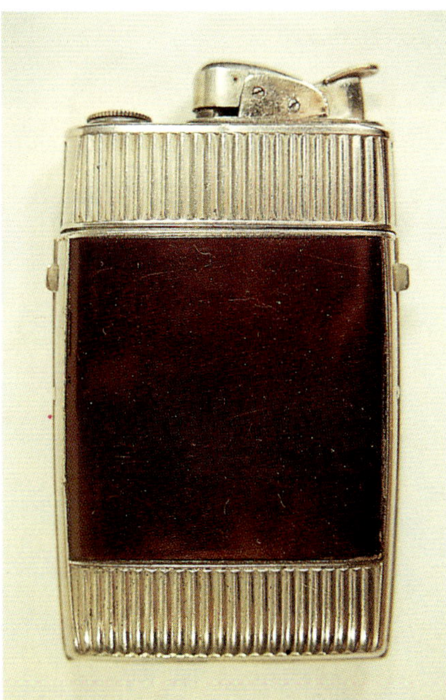

Chrome-plated and maroon enameled case-lighter in a ribbed design. $25-125.

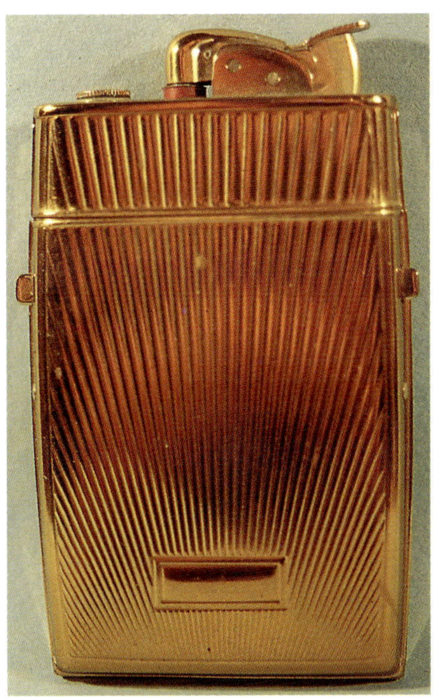

Gold-plated Monterey case-lighter with a sunray design. $25-60.

Trig-a-lite case-lighter in gold-plate and black enamel. $25-200.

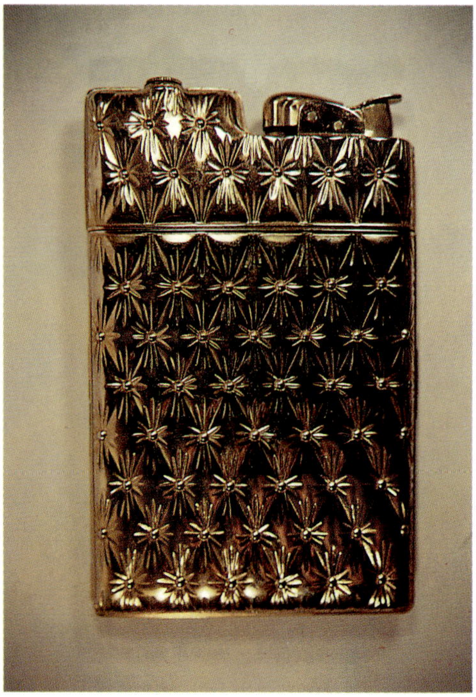

Gold-plated 20-Pak case-lighter in a floral pattern. $25-75.

Chrome-plated case-lighter with an art deco design. $20-50.

Black-enameled case-lighter with a Clipper mechanism and a built-in compact. $50-250.

Trig-a-lite case-lighter in yellow and brown enamel with a built-in compact. $40-225.

Trig-a-lite case-lighter in dark green enamel with a built-in compact. $40-225.

Black-enameled case-lighter with a white enamel compact lid in a floral design. $40-200.

Trig-a-lite case-lighter in black and cream enamel in an art deco design. $25-225.

Cavalier model case-lighter in chrome-plate with black enamel stripes. $25-50.

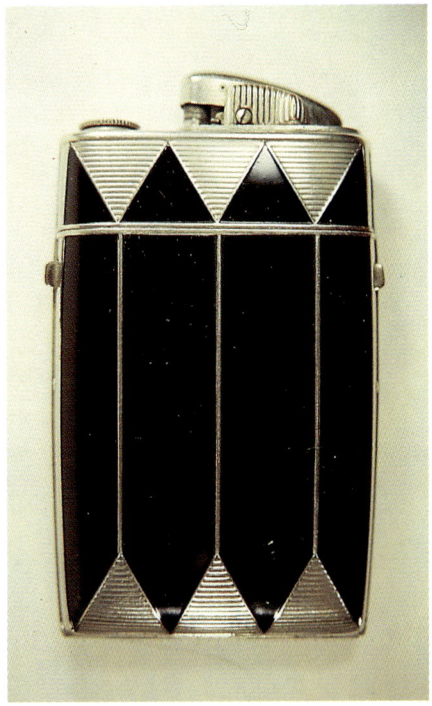

Case-lighter in chrome-plate and black enamel with a Clipper mechanism. $25-175.

Trig-a-lite case-lighter in maroon enamel. $25-150.

Trig-a-lite case-lighter in light green enamel with an art deco pattern. $25-225.

Chrome-plated case-lighter with a built-in watch. $60-175.

Trig-a-lite case-lighter in black and cream enamel with a built-in compact. $40-225.

Chrome-plated case-lighter in a herringbone/hobnail pattern with black enamel stripes. $25-75.

Trig-a-lite case-lighter in chrome-plate with a herringbone/hobnail pattern. $25-125.

Chapter 7
TABLE LIGHTERS

The Doulton/Evans Table Lighter

This table lighter was first produced by Royal Doulton in 1958. Evans inserts (and Colibri inserts) were installed in small size character jugs. Prior to the issue of character-jug lighters by Royal Doulton, another company (Van Cleff Studios) had issued character jugs with Evans inserts. As the Evans insert was smaller in diameter than the hole at the top of the these jugs, the hole diameter was reduced by the addition of plaster of Paris in order to make the insert fit. Van Cleff 's character-jug lighters portrayed some of the same characters as those portrayed by Royal Doulton.

The character-jug lighters made by Royal Doulton are listed below. Values listed were established by Royal Doulton collectors and do not necessarily reflect the item's value to a cigarette-lighter collector. On this and the following page are pictures of three character-jug lighters from the collection of Judith Sanders.

Name	Dates Made	Value
Bacchus	1964-1973	$260-395
Beefeater	1958-1973	$115-175
Buz Fuz	1958 only	$300-395
Captain Ahab	1964-1973	$250-595
Captain Cuttle	1958 only	$295-325
Falstaff	1958-1973	$125-200
Long John Silver	1958-1973	$90-120
Mr. Micawber	1958 only	$200-250
Mr. Pickwick	1958-1961	$225-350
Musketeer (Porthos)	1958 only	$425-475
Old Charley	1959-1973	$195-300
Poacher	1958-1973	$110-150
Rip Van Winkle	1958 only	$750-900
Lawyer	1960 only	$200-250
Sairey Gamp	unknown	$200-250 (est.)

Sairey Gamp. $200-250.

Buz Fuz. $300-395.

Falstaff. $125-200.

The Sascha Brastoff/Evans Table Lighter

Sasha Brastoff was a renowned and creative artist and ceramics designer. He began his career in the 1940s and continued producing innovative ceramic wares until his death in 1993. He worked in a Los Angeles studio built for him by his friend Winthrop Rockefeller.

During his lifetime, Brastoff had many occupations in addition to his work as a talented ceramics designer. He was a painter, sculptor, dancer, costume designer, and jeweler. His ceramic work was much admired by the rich and famous and his wares commanded much higher prices than those of other ceramic designers of the time.

Brastoff supervised a staff of over one hundred artisans who produced his ceramic wares in volume. Items created by him personally were signed "Sascha Brastoff." Items produced by his workers were signed "Sascha B." or were stamped with his full name.

Of interest to Evans lighter collectors is the fact that at some point in his career he became connected

Sascha Brastoff / Evans table lighter. $20-40.

to the Evans Case company, producing bases for Evans table lighters. These lighters are seldom seen, which is indicative of only modest production. Pictured here are three of Sascha Brastoff's table-lighter designs. The first two lighters pictured are signed "Sascha B." on the side. All three lighters are stamped on the bottom "Sascha Brastoff, California" in gold lettering along with a rooster design. The third picture is of a lighter and a cigarette holder with a matching floral design. As Brastoff also designed ceramic ashtrays, there is a good chance that some lucky lighter collector may have a three piece matching set.

Sascha Brastoff/Evans table lighter. $20-40.

Sascha Brastoff/Evans table lighter and matching cigarette holder. $40-75 set.

The Clearfloat/Evans Table Lighter

In the 1950s a company in Attleborough, Massachusetts created a table lighter in partnership with the Evans Case Company. This company, Clearfloat, Inc., was founded in 1946 by Roy Slipp and is still in business in Attleborough today, although it no longer produces table lighters. When Clearfloat was producing table lighters, it manufactured the bases and Evans Case manufactured the lighter mechanisms. The bases were made of clear acrylic that is trademarked under the name "CLEARFLOAT." When viewed from the bottom while held up to a light, the name "CLEARFLOAT" can be seen in the acrylic on lighters that were made to be sold by Clearfloat, Inc. Lighters that were made to be sold by Evans Case do not have the "CLEARFLOAT" name. Three Clearfloat table lighters belonging to OTLS members are pictured in this section; additional Clearfloat table lighters are also pictured throughout this chapter. Instructions that came with Clearfloat table lighters read:

> This gem-like crystal-clear block is made by a remarkable new process, a product of Clearfloat, Inc., Attleboro, Mass. Created in an area which produces most of America's fine jewelry, the "CLEARFLOAT" process is the accomplishment of a combination of highly skilled artisians [sic] working hand-in-hand with expert technicians. This piece was custom built for your use and pleasure. If the surface becomes dulled in time, its original lustre may be restored by Simonizing. It was designed to serve you a lifetime.

Today, Clearfloat, Inc. is solely owned by Tim Murphy. The company currently manufactures trophies, awards, and executive gifts. Many are made of clear acrylic. The Hilsinger Company is one of Clearfloat's customers. Clearfloat, Inc. provides Hilsinger with display items for trade show displays created by Hilsinger. Also, as of this writing, Clearfloat is encapsulating small meteorites in CLEARFLOAT for NASA which will be used for NASA presentations in schools.

Clearfloat/Evans table lighter advertising Chesterfield Cigarettes. $15-35.

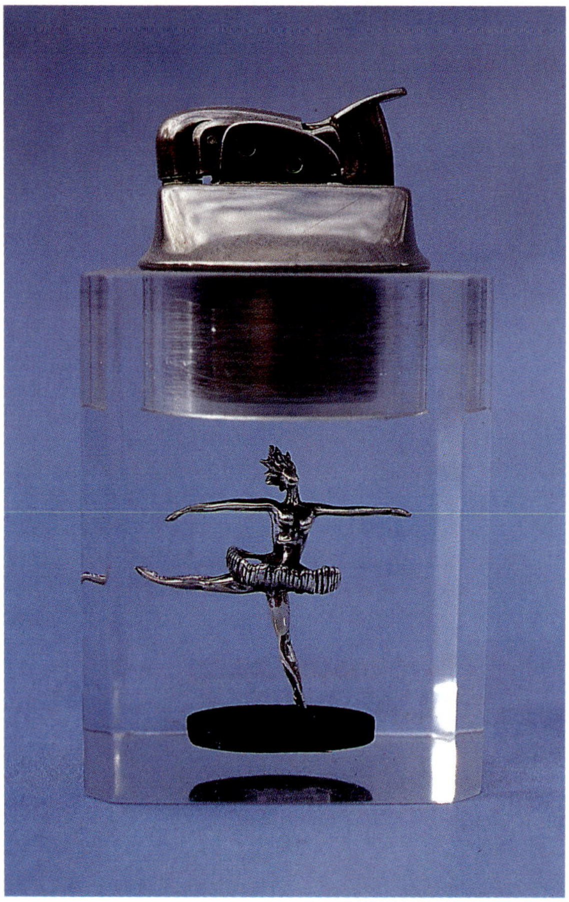

Clearfloat/Evans table lighter named the "Ballet Queen." $15-35.

Clearfloat/Evans table lighter showing U.S. coins. $15-35.

King and Queen white ceramic chess piece table lighters. $25-45 each.

Small table lighter in green onyx. $15-30.

White cloisonné-enameled bell-shaped table lighter. $50-125.

Three piece frosted glass set consisting of a cigarette lighter, cigarette holder, and ashtray. $75-150 set.

Left:
Trig-a-lite table lighter with a green acrylic base. $25-100.

Windsor model table lighter with a golden metal finish. $15-40.

Gold-plated Magic Lamp model table lighter. $20-60.

Silver-plated Silver Queen model table lighter. $20-50.

Tall gold-plated table lighter in green enamel with a rooster design. $20-40.

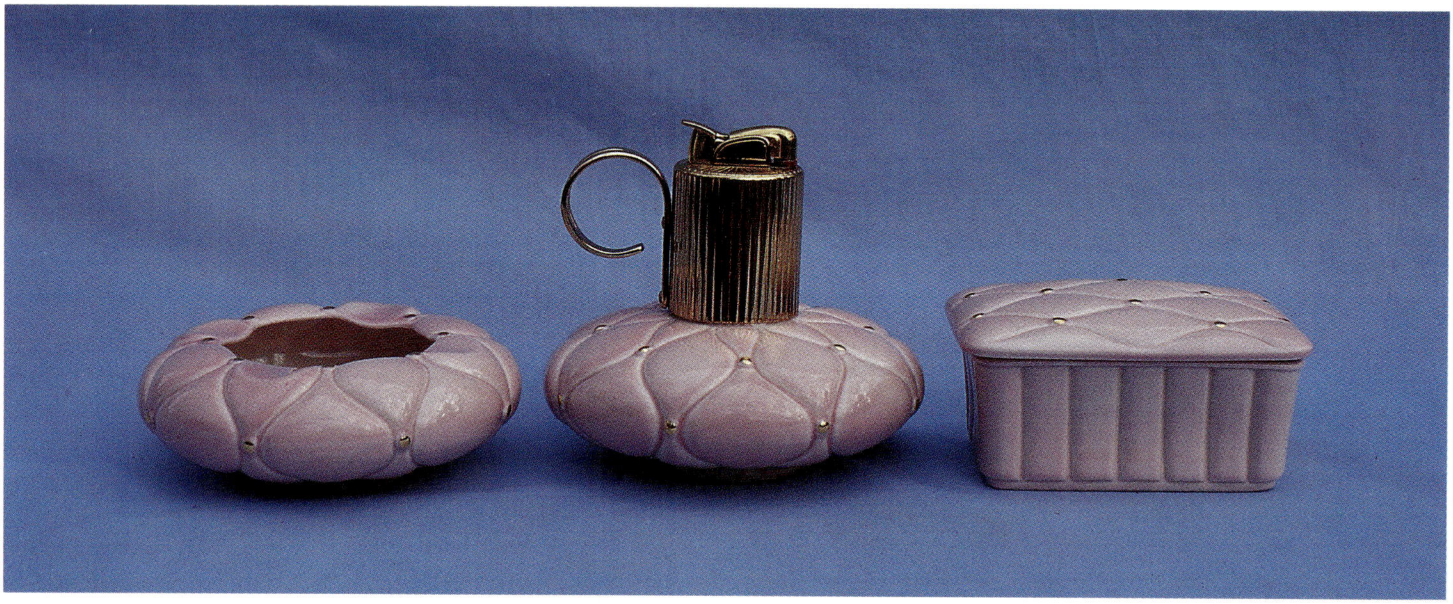

Three-piece table set in a pink quilted design. $50-100 set.

Sterling silver table lighter marked "J.E. Caldwell, sterling, weighted 555." $50-100.

Cornucopia model table lighter in silver-plate. $40-75.

Three-piece antique car table set. $40-75 set.

Chrome-plated table lighter with a red and black flag. $15-40.

Two-piece Trojan warrior table set in grey and black enamel. $20-50 set.

Table lighter in the shape of a brass drum. $20-40.

Ceramic table lighter in the shape of a set of books. $15-40.

Clearfloat/Evans table lighter advertising Fruehauf Volume Van. $15-35.

Brown ceramic table lighter. $15-40.

Two-piece table set in a deep blue enamel. $50-100 set.

Shadowbox design table set showing a Currier & Ives print. $20-50 set.

Gold-plated, moss green table lighter on a pedestal. $25-50.

An unusual gold-plated, double-ended table lighter in brown acrylic. $50-100.

A tall, cut-glass table lighter. $15-40.

A beautiful table lighter in brass and milk-white glass. $20-40.

Gold-plated table lighter with brandy-colored enamel. $25-50.

Chrome-plated Trig-a-lite table lighter in an art deco design. $40-100.

Table lighter in brass and enamel in the shape of a world globe. $20-40.

Table lighter with a clear acrylic base and a Clipper mechanism. $25-50.

Silver-plated table lighter in the shape of a trophy. $20-40.

Table lighter in brandy-colored enamel with a flared thumb piece. $25-50.

Chrome-plated Clearfloat/Evans table lighter with flowers in the base. $15-35.

Gold-plated table lighter in fine bone china. Circa 1955. $20-40.

Gold-plated table lighter with a wooden base and a flared thumb piece. $15-40.

Tall, moss green enameled table lighter on a silver-plated base. $50-100.

Table lighter with a white glass base and gold paint design. $25-45.

Gold-plated table lighter with a leather-covered, box-shaped base. $15-40.

Gold-plated table lighter with a beautiful blue, white, and gold enameled pattern. $25-75.

Gold-plated table lighter with a yellow cloisonné enamel base. $25-75.

Clearfloat/Evans table lighter in chrome-plate with a flying pheasant. $15-40.

Table lighter in heavy sterling silver. $75-150.

Leather-covered cigarette box with built-in lighter. $20-50.

Small gold-plated table lighter with a black snakeskin covering. $15-40.

Clearfloat/Evans table lighter in gold-plate with a miniature covered wagon. $15-40.

Silver-plated Duchess model table lighter. $15-40.

Glass globe table lighter with a miniature Christmas Tree. $15-30.

Small gold-plated table lighter with a glass base. This lighter is from the collection of Judith Sanders and once belonged to her grandmother. $15-30.

Tall gold-plated table lighter with a blue enamel base. $20-60.

Gold-plated Clearfloat/Evans table lighter with flowers in the base. $15-40.

Table lighter in gold-plate with a light yellow enamel base. $25-75.

A beautiful, dark green three-piece table set in gold-plate. $50-100 set.

Gold-plated table lighter with a light green cloisonné enamel base. $20-50.

Table lighter and ashtrays in a Wedgwood pattern. $40-100 set.

Small table lighter on a miniature carriage. $15-40 set.

Clearfloat/Evans table lighter with a jeweled-crown design. $15-40.

Gold-plated Phinney-Walker clock table lighter. $20-75.

Gold-plated table lighter with an unusual base. $15-40.

Seven-piece table set in gold-plate. $100-200 set.

Table lighter in gold-plate with a light yellow cloisonné enamel base. $25-75.

Gold-plated moss green table lighter with matching ashtray/candy dish. $50-125 set.

Tall gold-plated table lighter with Triangle logo. $20-40.

Tall Roller-bearing table lighter in silver-plate with an art deco design. Circa 1930. $50-100.

Gold-plated table lighter with a glass base. $15-40.

Brass and cut-glass table lighter. $20-50.

Below:
Gold-plated table lighter with a miniature cannon for a base. $35-75.

White ceramic horse's head table lighter with a chrome-plated base. $20-75.

Gold-plated table lighter with a white-enameled, gold-painted base. $20-40.

Gold-plated table lighter with a moss green cloisonné enamel base. $25-75.

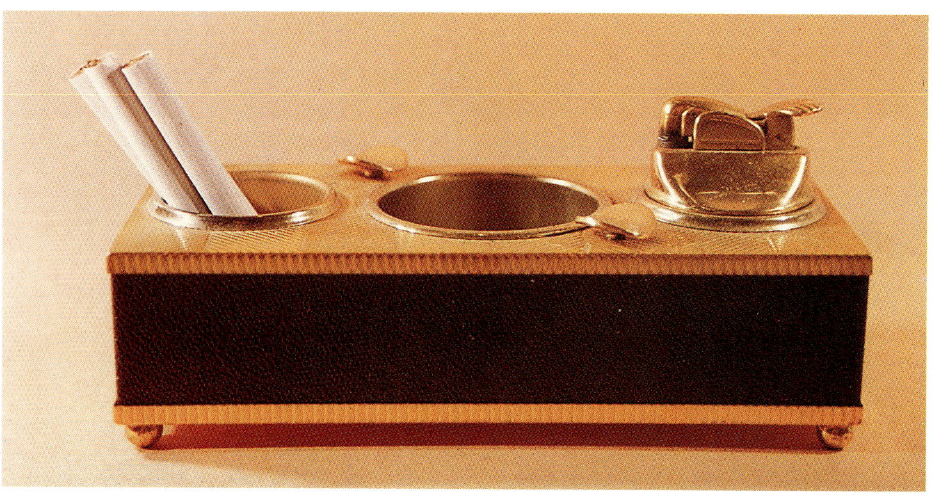

Executive desk set in black Moroccan leather. $20-50.

Chrome-plated Clearfloat/Evans table lighter. $20-40.

Gold-plated table lighter with a hexagonal green and brown onyx base. $20-40.

Gold-plated leather-covered table lighter advertising Carolina Garage, Inc. $20-40.

Gold-plated table lighter with a red and white Limoges china base. $20-40.

Gold-plated table lighter with a light blue globe-shaped base. $25-50.

Gold-plated Flip-Top model table lighter in light and dark green, circa 1955. Table lighters of this design are seldom seen. $60-125.

Cloisonné-enameled table lighter and matching ashtray/candy dish, both in light yellow. $75-150 set.

Gold-plated table lighter in a white ceramic crown design. $20-50.

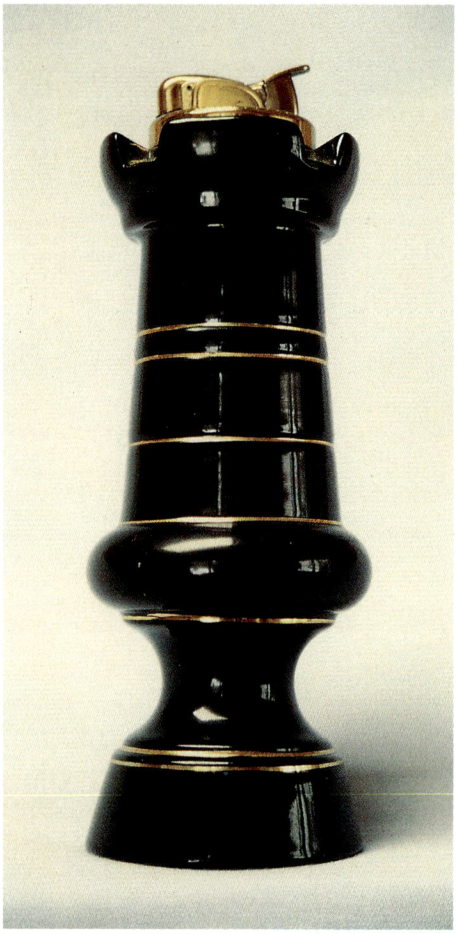

Black ceramic castle chess piece with gold-painted trim. $25-50.

Black ceramic horse's head with gold-plated base. $25-50.

Silver-plated table lighter and cigarette urn in the shape of a candelabrum. $50-100.

Small brass table lighter with a chrome-plated insert. $20-40.

Table lighter in brass and cranberry glass. $20-40.

Trig-a-lite table lighter picturing three horses. The top of the lighter lifts up to access a compartment that holds cigarettes. $75-175.

Clearfloat/Evans table lighter with a seabed scene. $15-40.

Gray ceramic table lighter, gold-plated, in the shape of a pitcher. $25-50.

Bell-shaped table lighter in brass and cobalt blue enamel. $50-125.

Heavy, sterling silver table lighter marked "Black Starr Frost, sterling, weighted, 888." $75-150.

Tall gold-plated table lighter in grey and black enamel showing a Trojan warrior and sword. $20-40.

Silver-plated Magic Lamp table lighter with original tags. $25-75.

Tall table lighter in the shape of a woman carrying a basket. Imitation rubies are strategically placed. $100-200.

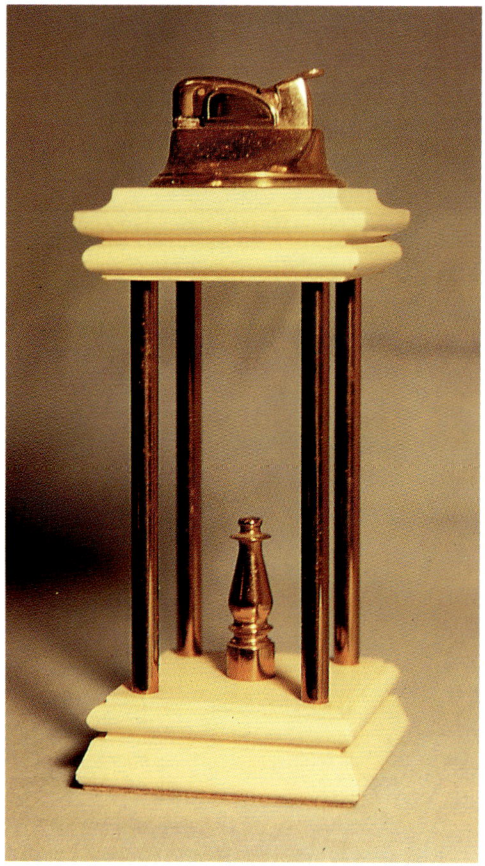

Tall gold-plated table lighter with column base. $25-50.

Chrome-plated Roller-bearing table lighter. $25-50.

Table lighter in heavy sterling silver. Very similar in design to a lighter which is named the "Princess" and shown in an Evans catalog. $50-100.

Gold-plated table lighter in the shape of a beanbag with a plaid fabric design. $20-40.

Tall table lighter with a silver-plated brass and wood base in the shape of a hunting horn. Similar to the Dunhill Hunting Horn table lighter. $50-100.

Gold-plated table lighter with a brass base. $25-50.

Table lighter with a wooden base in the shape of a horse's backside. $25-40.

Table lighter listed as the "Classic." Cobalt blue enamel and gold-plate finish. $25-50.

Chrome-plated lift-arm table lighter in the shape of a teapot, including handle and lid. Evans lift-arm table lighters are seldom seen. $75-150.

Tall table lighter with a glass base that has a gold-paint trim. $25-75.

Three-piece table set with lighter, cigarette urn, and ashtray/candy dish. The ashtray/candy dish is listed as the "Versa-tray."

Left:
Bell-shaped table lighter in light yellow cloisonné enamel. $50-125.

Table lighter with a fancy ceramic base and the name "Madelaine." $20-40.

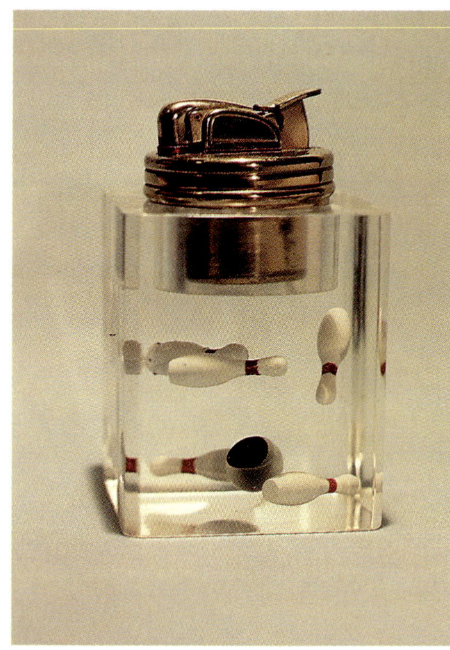

Clearfloat/Evans table lighter with embedded miniature bowling ball and pins. $15-40.

Table lighter with a white ceramic base and a leaf design. $15-35.

Tall table lighter showing Grecian women in flowing dresses. $20-40.

Gold-plated table lighter with a wooden base in the shape of a block. $20-40.

Lift-arm table lighter with a wooden base. $50-150.

Gold-plated table lighter with a horse and jockey and a wooden base. $50-100.

Black ceramic king and queen chess pieces. $25-50 each.

Gold-plated table lighter in the shape of a western boot with spur. $25-50.

Table lighter with a wooden base in the shape of a teapot. $20-40.

A grouping of king, queen, and castle chess pieces. $25-50 each.

White ceramic table lighter with gold-paint trim. Shaped like a lamp. $25-50.

Three-piece table set in gold-plate. $100-200 set.

Stein lighter with sculpted man and woman dancing. $15-40.

Gold-plated table lighter with a base of brass and moss green enamel. $25-75.

Clearfloat/Evans table lighter in hexagonal shape. The design is called "Sea life" in a Clearfloat catalog. $15-40.

Clearfloat/Evans table lighter with an embedded red rose. $15-40.

An unusual lighter with a seashell design embedded in green acrylic. This may be a Clearfloat/Evans lighter. $50-100.

Wooden table lighter/cigarette dispenser called the "Pagoda" by Evans collectors. When it is wound up it plays music. When the music stops, little doors open to dispense cigarettes. $100-200.

Pagoda lighter in the doors-open position.

Unusual table lighter in brown leather, shaped like a basket with a handle. $20-40.

Table lighter similar to one pictured previously. This one is in chrome-plate. $15-40.

Trig-a-lite table lighter in black enamel and chrome. This lighter also hold cigarettes. $75-150.

Clearfloat/Evans table lighter named "Lighthouse and Gulls" in Clearfloat's catalog from the 1950s. $15-40.

Trig-a-lite table lighter with matching cigarette container. Both depict scenes from a fox hunt. $75-150 set.

Left:
Tall gold-plated table lighter on a pedestal. $40-75.

Green marble or onyx table lighter with matching ashtray. $50-100 set.

Clearfloat/Evans table lighter called "The Duffer" because the golfer is in a sand trap. $15-40.

Small ceramic table lighter with a yellow base and floral design. $15-40.

Man smoking pipe. Brass base with gold-plated insert. $25-50.

Silver-plated table lighter with gray enamel base. $20-60.

Four-piece table set in moss green enamel. $75-150 set.

Clearfloat/Evans table lighter with a brown and white running dog. $15-40.

Gold-plated, white cloisonné-enameled table lighter with matching ashtray/candy dish. $75-150 set.

A silver-plated table lighter made by International Silver Plate with an Evans insert. Part of a set that includes a cigarette container. $50-100 set.

Cigarette container which is part of the set that includes the lighter pictured at left.

Table lighter in the shape of a lipstick tube. White enamel and gold-plated base with a gold-plated lid. Circa 1955. $40-75.

Gold-plated table lighter with a base of brass and moss green cloisonné enamel. $40-75.

Gold-plated table lighter with a base made of an early global map. $25-50.

Ceramic table lighter showing a man with a sad, comical face. $25-50.

Gold-plated table lighter with a base of brass and blue glass. $40-75.

Clearfloat/Evans table lighter showing a fish striking a fishing fly. $15-40.

Small gold-plated, leather-wrapped table lighter. $15-30.

Clearfloat/Evans table lighter with the "Sea life" design. $15-40.

Cigar store Indian with full paint. This lighter has also been seen without paint. $100-200.

Cigar store Indian (side view).

Gold-plated table lighter with a colorful ceramic base. $40-75.

Large gold-plated table lighter in the shape of a heart with a basket-weave pattern. $20-40.

Silver-plated Trig-a-lite table lighter in black enamel picturing a dog. $50-175.

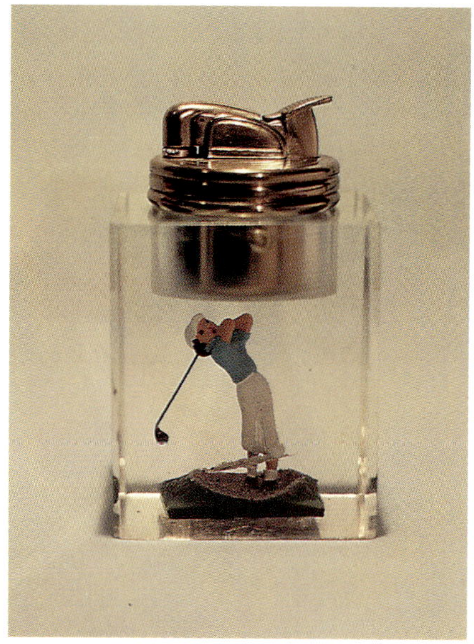

Clearfloat/Evans table lighter, "The Pro Golfer." $15-40.

Gold-plated table lighter with a moss green enamel base. $40-75.

Musical, revolving, smiling "Top Kick" table lighter. Well known to Evans collectors. Awarded as a prize in a punch-board game. $20-40.

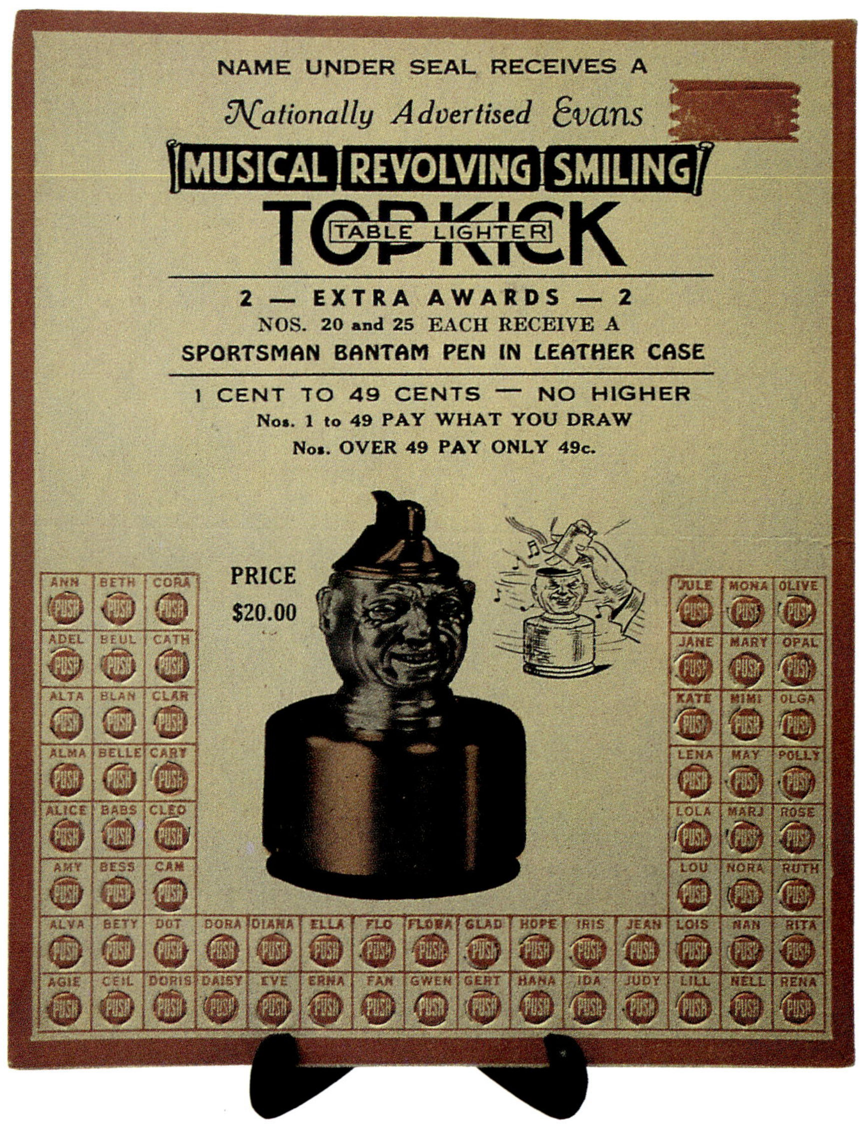

Punch-board for "Top Kick" table lighter.

Gold-plated table lighter with a dark blue enamel base. $40-75.

Three-piece brandy-colored enameled table set. $75-150 set.

Two white ceramic castle chess pieces. $25-50 each.

Tall gold-plated table lighter with advertising labels. $20-40.

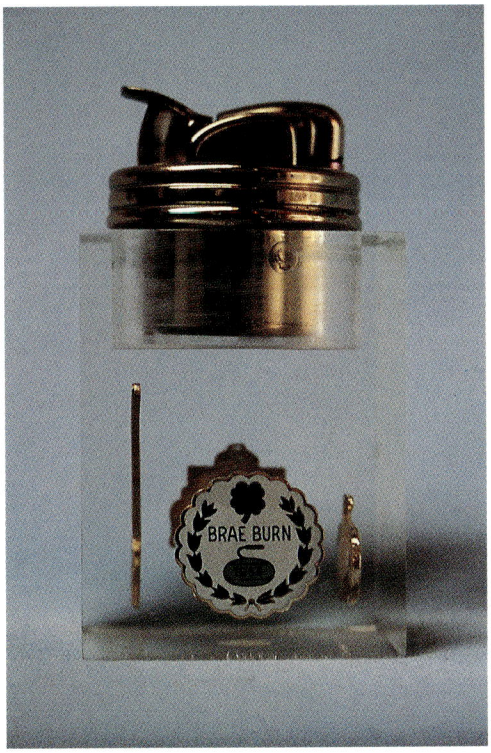

Clearfloat/Evans table lighter advertising "Brae Burn, 1955." Brae Burn may have been a country club. $15-40.

Two-piece table set in white cloisonné enamel with a floral pattern. $75-150 set.

Pagoda table lighter in varnished wood. $100-200.

Pagoda table lighter in open position.

Clearfloat/Evans table lighter with red and green fishing flies. $15-40.

Executive desk set in gold-plate and black Moroccan leather. $20-40.

Clearfloat/Evans table lighter with an embedded "Playboy" key. $15-25.

Combination table lighter and bowling trophy in gold-plate. $50-100.

Table lighter with a jockey in red and white colors. $25-50.

Chrome-plated Trig-a-lite table lighter with a green and yellow onyx base. $50-100.

A nude lady in brass with her legs balancing a gold-plated table lighter. $100-150.

Below:
Desk set with a lighter, cigarette holder, and golfer in gold-plate. $75-150.

Gold-plated table lighter with a ceramic base in a bird pattern. $20-40.

Gold-plated table lighter on the back of a miner's donkey. Note the real hair on the donkey's tail. $40-75.

Green glass table lighter with a sterling silver overlay. $75-150.

Gold-plated table lighter in a top hat made of Limoges china. $40-75.

Five-piece table set made of Herend china in the Sang Jaune pattern. A set of china in this pattern was given by Emperor Franz-Joseph to his wife, Elizabeth. $200-300 set.

Three-piece table set in chrome-plate and red Bakelite. $50-150 set.

Chapter 8

FRUIT AND EGG LIGHTERS

Left:
Acorn lighter made of cut glass with a brass base and trim. $40-75.

Right:
Large vertical egg in white cloisonné enamel with gold-plated base and trim. $75-150.

Blue enameled pear with matching ashtray. $60-125 set.

White Imperial egg in the Fabergé style. $50-125.

Pineapple with a satin finish. $35-100.

Strawberry in white enamel with brass base and trim. Enameled strawberries are difficult to find. $75-150.

Egg in red cloisonné enamel with brass base and trim. This egg is shaped a little differently than eggs pictured earlier. $50-125.

Small horizontal egg in dark green enamel with brass base and trim. $50-125.

Large vertical egg in moss green enamel with gold-plated base and trim. $50-125.

Two-tone green enameled pear with matching ashtray/candy dish. $75-125 set.

White cloisonné-enameled pear with gold-plated base and trim. $75-150.

Vertical egg and matching ashtray/ candy dish, both in light yellow cloisonné enamel. $100-175 set.

Pineapple in cut-glass with brass base and trim. $40-75.

Vertical egg with a mesh top and a lavender enameled base. $75-150.

Pear in moss green enamel with gold-plated base and trim. $50-125.

Copper-colored pear with matching ashtray/candy dish. $60-125 set.

Imperial egg in turquoise cloisonné enamel with gold-plated base and trim. $50-125.

The lighter at left in the open position.

Pineapple in light yellow enamel. $50-125.

Horizontal egg in brushed gold-plate. $40-100.

Brass pear. $40-100.

Pineapple in green enamel. This pineapple does not have a base. $50-100.

Vertical egg in cobalt blue with a tall base. $50-125.

Tall wine-colored enameled egg with matching ashtray/candy dish. $75-150 set.

Vertical egg in light pink cloisonné enamel with gold-plated base and trim. $75-150.

Left:
Pineapple in light blue with gold-plated trim. $50-125.

Imperial egg in white cloisonné enamel. $50-125.

Pineapple in black glass with brass trim. $40-75.

Left:
Brandy-colored vertical egg with matching ashtray. $75-150 set.

Pear in light yellow cloisonné enamel with brass trim. $50-125.

Horizontal egg in brushed gold-plate. $40-100.

Apple in a satin finish with gold-plated trim. $40-100.

Apple in green cloisonné enamel. $50-100.

Apple with a copper-colored hammered finish. $40-100.

Vertical egg in moss green enamel with candlesticks. Lighter: $50-125. Candlesticks: $40-75 set.

Apple in dark green enamel with gold-plated trim. $50-100.

Vertical egg in brass with an eagle on the top. $40-75.

Left:
Horizontal egg with a mesh top and bottom. $50-100.

Imperial egg in blue cloisonné enamel. $50-125.

Acorn in green enamel with brass trim. $40-100.

Acorn in reddish-brown enamel with gold-plated base and trim. $50-125.

Acorn in light yellow enamel with brass trim. $40-100.

Vertical egg in cobalt blue with matching ashtray/candy dish. $75-150 set.

Pineapple in white enamel with gold-plated trim. $50-125.

Apple with a brushed gold-plate finish. $40-100.

Pineapple in brandy-colored enamel with gold-plated trim. $50-125.

Imperial egg in light pink cloisonné enamel with a small decoration on the top. $50-125.

Acorn with a wooden top and a brass base. $40-100.

Apple in wood with gold-plated trim. Fruit made of wood is harder to find than enameled fruit. $40-100.

A four-piece set consisting of an egg lighter, cigarette holder, ashtray, and candy dish. $200-300 set.

An unusual gold-plated vertical egg with a decorative base. $100-200.

Imperial egg in dark blue cloisonné enamel with gold-plated base and trim. $50-125.

Apple with a brushed gold-plated finish. $40-100.

Small pineapple in blue enamel. $50-125.

Vertical wooden egg with gold-plated base and trim. $40-100.

The same egg in the open position.

Yellow cloisonné-enameled Imperial egg with brass base and trim. $50-125.

Apple in cut-glass with gold-plated trim. $40-75.

Blue enameled pear with silver-colored trim. $40-100.

Imperial egg in green enamel with brass base and trim. $50-125.

Pineapple in dark green enamel with gold-plated base and trim. $50-125.

Vertical egg in blue enamel with gold-plated base and trim. $50-125.

Melon in blue enamel with brass base and trim. $50-100.

Acorn with green top and brown bottom. $40-100.

Apple in yellow cloisonné enamel with gold-plated trim. $50-125.

Light yellow enameled apple with matching ashtray/candy dish. $75-150 set.

Apple in dark blue enamel with gold-plated trim. $50-125.

Imperial egg in yellow cloisonné enamel with gold-plated base and trim. $50-125.

Enameled pear with matching ashtray/candy dish. $75-150 set.

Brandy-colored Imperial egg with brass base and trim. $50-125.

Red cloisonné-enameled Imperial egg with matching candy dish. $75-150 set.

Pear in light yellow cloisonné enamel with gold-plated trim. $50-125.

Vertical egg with brass base and trim. $50-125.

Acorn in green enamel with matching ashtray/candy dish. $75-150 set.

Imperial egg in light yellow cloisonné enamel. Notice the stripe pattern in the enamel. $50-125.

Pear in cut-glass with gold-plated brass trim. $40-100.

Left:
Pear in black enamel with gold-plated trim. Black enamel is the hardest color enamel to find. 450-125.

Vertical egg in cobalt blue cloisonné enamel with gold-plated base and trim. $50-125.

Acorn in brass. $40-100.

Tall vertical egg in green cloisonné enamel with matching ashtray/candy dish. $75-150 set.

Pineapple in gold-plated brass. $40-100.

Small vertical egg in turquoise cloisonné enamel with matching ashtray/candy dish. $75-150 set.

Vertical egg in light blue cloisonné enamel with gold-plated base and trim. $50-125.

Melon in green enamel with brass trim. $40-100.

Pineapple in brandy-colored enamel with gold-plated trim. $50-125.

Vertical egg in red cloisonné enamel with gold-plated brass trim.

Pineapple in red enamel with gold-plated trim. $50-125.b

A group of enameled apples in a variety of colors. $50-125 each.

A group of enameled pears in a variety of colors. $50-125 each.

Vertical egg in dark blue enamel with gold-plated base and trim. $50-125.

A group of enameled fruit and eggs. $50-125 each.

A group consisting of an enameled egg and various enameled table lighters. $40-125 each.

A group consisting of enameled eggs and table lighters. $40-125 each.

A group of cloisonné-enameled Imperial eggs. $50-125 each.

Left:
Two vertical cobalt blue enameled eggs and a matching cigarette box. Lighters: $50-125 each. Cigarette box: $25-50.

An enameled table lighter and a set of candlesticks. Lighter: $40-60. Candlesticks: $50-75 set.

Chapter 9
Display Cases and Other Items

Display case containing mint compacts and lighters. $500 for the case and contents.

Gift set of mint table lighters and ashtrays. $350 set.

Lower level of the display case above.

Display case of mint pocket and case-lighters. $300.

Wooden display case without the lighters. $100.

Display case of mint pocket lighters. $300.

White enamel cigarette holder with a bee pattern. $25-50.

Wooden display case with mint lighters and a glass front. $350.

Miss Evans Little Shaver. $20-50.

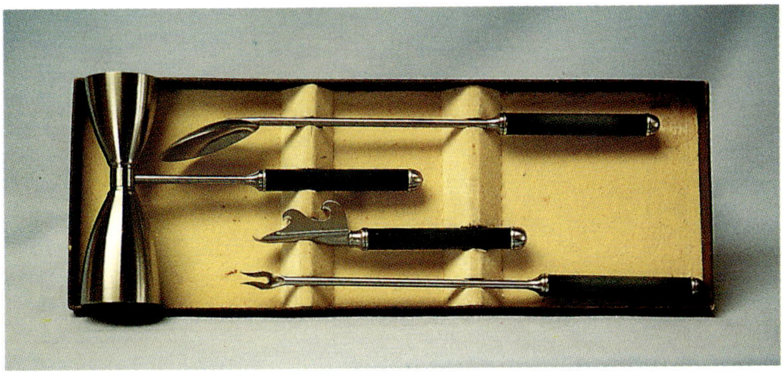

Bar set made by Evans Case. $20-50.

Vanity mirror in gold-plate and light green enamel with a floral design. $40-75.

Vanity mirror in gold-plate with an engraved design. $40-75.

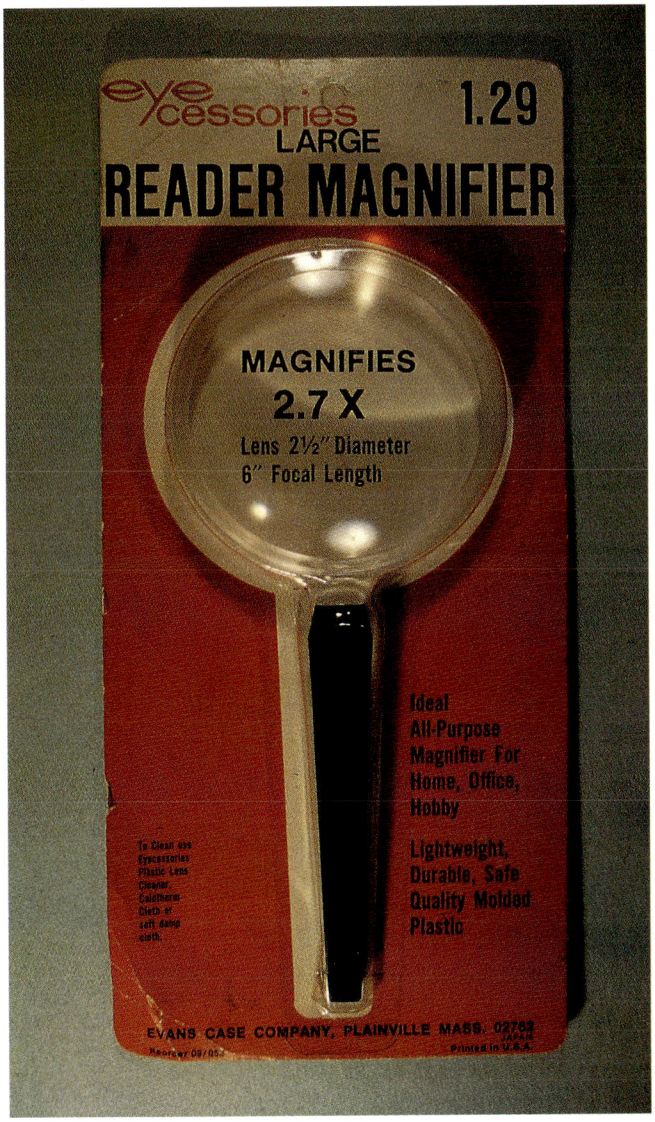

Large reader magnifier, marked "Evans Case Company, Plainville, Mass." Actually made by Hilsinger. $25.

Combination cigarette case and compact. White and black enamel in an art deco pattern. $25-75.

Cigarette case in gold-plate with a ruby finish. $50-100.

Jewelry box in yellow cloisonné enamel. $50-100.

Rosary holder with a satin enamel finish and gold-plated base and trim. $75-150.

Candy dish in a ruby finish with gold-plated trim. $25-50.

Silver-plated salt and pepper shakers. One of Evans Case's first products. $50-100 set.

Combination table lighter and lamp called the "Lamp-lighter." $100-200.

Chapter 10

Compacts, Perfumers, and Handbags

Rectangular compact in gold and silver-plate. A beautiful item. $60-90.

Gold-plated compact with a geometric pattern. $35-50.

Compact and matching lipstick holder. Both in a ruby finish with a fleur-de-lis pattern. $60-100.

Small compact in white enamel with a floral design. $25-45.

Compact in black enamel and white cloisonné enamel with a floral design. $40-75.

Small compact with a gold-colored mesh bottom. $50-90.

Compact with a simulated jewel decoration. $35-60.

Oval compact in highly-polished chrome. $30-60.

Small compact with an attached lipstick holder. Both in a ruby finish with a fleur-de-lis design. $75-125.

Square compact in silver-plate with a floral design. $40-60.

Compact in white enamel and highly-polished chrome. $50-75.

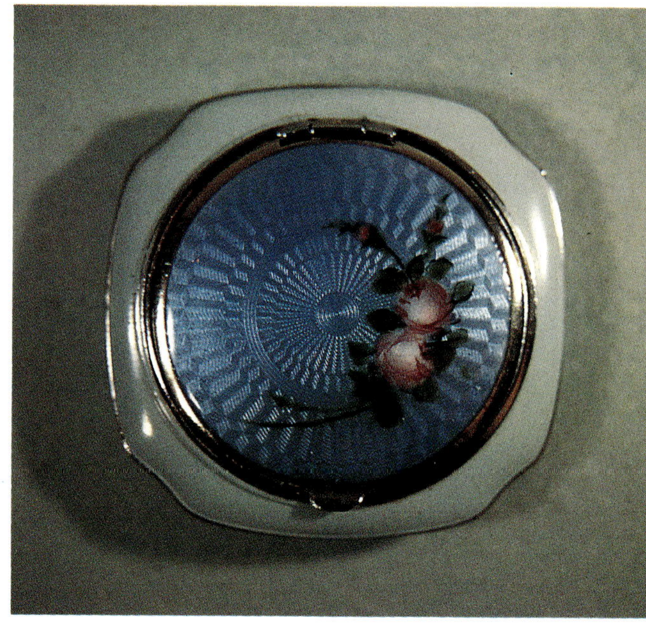

Compact in white enamel and blue cloisonné enamel with a floral design. $35-60.

Compact in gold-plate and mother-of-pearl. $40-60.

Gold-plated rectangular compact with an engraved design. $45-70.

Left:
Oval compact in white enamel with a floral design. Called the "Coronet." $40-60.

Right:
Engraved metal compact with a floral design. $35-60.

Compact in white enamel with simulated pearl decoration. $35-60.

Compact in chrome-plate and black enamel. $40-60.

Compact in green cloisonné enamel with a floral pattern. $75-100.

Gold-plated engraved compact. $25-45.

Square compact in blue cloisonné enamel with a floral design. $45-65.

Compact in gold-plate with emerald-colored stones. $75-100.

Compact in gold-plate with a floral design in white cloisonné enamel. $50-80.

Compact in blue cloisonné enamel with a floral design. $75-100.

Compact in rhodium-plate with a sun-burst pattern. $150-200.

Left: Musical compact with blue stones. $60-90. Right: Musical compact with mother-of-pearl. $60-90.

Pendant vanity with an engine-turned pattern. $100-150.

Compact in highly-polished chrome with a swirl pattern. $35-60.

Square compact in white enamel with lavender flowers. $30-50.

Compact in white cloisonné enamel with a red rose. Named the "Nosegay." $75-100.

Compact in white enamel with flowers. $25-45.

Gold-plated watch compact in a sunburst pattern. $150-200.

Compact in black and white enamel with a flower basket. $35-60.

Compact in gold-plate and brown enamel. $35-60.

Gold-plated mother-of-pearl compact with wrist chain. $50-75.

Compact in yellow cloisonné enamel with a gold-colored mesh base. $75-100.

Cigarette lighter with a compact for a base. This item cannot function as a lighter nor as a compact and is thought to have been made by a factory worker who was looking for something to keep himself or herself busy. It certainly did not get to the Quality Assurance Department. $50.

White enameled compact with a green-colored stone. $35-60.

Perfumer in white cloisonné enamel and an art deco pattern. $150-200.

Compact in white cloisonné enamel with a rose pattern. $75-100.

Perfumer in gold-plate and green enamel with a flower basket. $100-200.

Table lighter with matching table perfumer. Made by Evans Atomizers, Inc., Brooklyn, New York. $40-75 set.

Egg perfumer in blue enamel with a gold-plated base and trim. $75-150.

Perfumer in chrome-plate with a multicolored art deco design. $100-200.

Perfumer in blue cloisonné enamel with a floral pattern. $150-200.

Perfumer in yellow and brown enamel in an art deco design. $150-200.

Handbag in white leather with a crown emblem. $75-125.

Perfumer in yellow cloisonné enamel with a man serenading two women. $200-300.

Handbag in red leather with accessories. $100-250 set.

Handbag in black leather with accessories. $100-200 set.

Com-purse in gold-colored mesh. The com-purse is a combination compact and purse. The compact is built into the underside of the purse lid and has a compartment for powder and rouge. $100-200.

Com-purse in gold-colored mesh, lid of blue cloisonné enamel with a floral pattern. $100-200.

Carryall in a rhodium finish with a sunburst design. Named the "Coronation." $75-150.

Gold-plated Carryall with its original box. $75-150.

Gold-plated Carryall with diagonal lines. $75-150.

Gold-plated Carryall in a basket weave pattern in original gift box. The box was intended to be used as a jewelry box after removal of the Carryall. $75-150.

Carryall in a rectangular shape with a basket weave pattern. $75-150.

The same carryall in the open position showing a powder compartment, powder puff, lipstick holder, change holder, and a comb.

Square Carryall with a basket weave pattern. $75-150.

The same Carryall with one side open. This side contains a lighter and cigarette holder.

Now the vanity side of the same Carryall is open showing its contents.

Chapter 11

The Dodge La Femme

The late 1950s was a time in America when style was everything. Automobile styling was no exception and Detroit learned that eye-catching styling was the way to sell cars. Automobile manufacturers recognized the importance of women as the dominant influence in new-car purchases. The automobile designers studied the color and design of fashionable women's clothing of the time and endeavored to color coordinate and pattern automobile exteriors and interiors to match the ladies' fashionable garments. Some automobile interiors were designed by women for women with such items as seasonally changeable seat covers. Designers such as Hubert de Givenchy, Elizabeth Arden, and Jacques Fath created gowns to match particular Cadillac model colors.

In January of 1955, Dodge introduced the "La Femme" concept, a Custom Royal Lancer Hardtop, with an exterior color of Heather Rose over Sapphire White. A gold "La Femme" nameplate on each fender replaced the Royal Lancer nameplate. The interior consisted of Heather Rose Cordigrain bolster and trim. The car was specifically designed for the fashionable woman. It included a shoulder bag in soft rose leather with a compact, lipstick, comb, coin purse, cigarette case, and lighter, fitted in a special compartment attached to the back of the passenger's seat. The shoulder bag and its contents were manufactured by the Evans Case company. Additionally, a rain cape, hat, umbrella, and sleeve were fitted into a compartment located on the back of the driver's seat. These items may also have been manufactured by Evans Case. The lining of both compartments was of Heather Rose Cordigrain to match the automobile's interior. Pictured here is a tall, pink, leather-wrapped Evans Supreme pocket lighter which may at one time have been one of the accessories in the shoulder bag. The lighter is presently in the collection of Evans collector Ted Wayland.

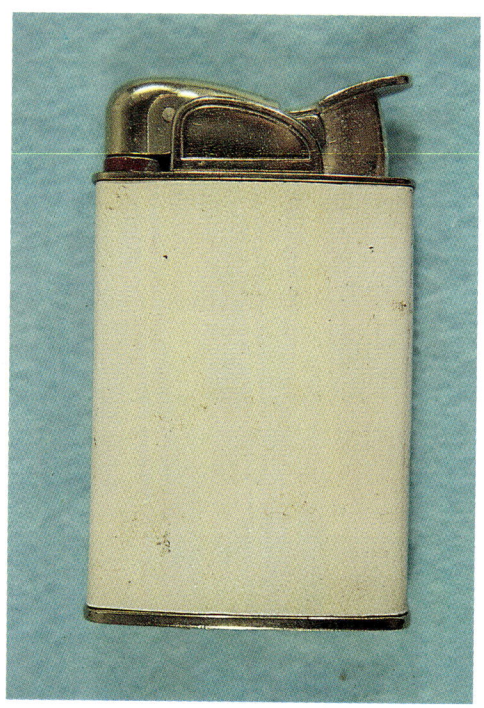

Supreme pocket lighter from the La Femme shoulder bag. *Collection of Ted Wayland*. $20-40.

Appendix
PATENT DRAWINGS

Patented June 28, 1927. Des. 72,965

UNITED STATES PATENT OFFICE.

ALFRED F. REILLY, OF ATTLEBORO, MASSACHUSETTS, ASSIGNOR TO EVANS CASE COMPANY, OF NORTH ATTLEBORO, MASSACHUSETTS, A CORPORATION OF MASSACHUSETTS.

DESIGN FOR A POCKET LIGHTER.

Application filed March 9, 1927. Serial No. 21,071. Term of patent 7 years.

To all whom it may concern:

Be it known that I, ALFRED F. REILLY, a citizen of the United States, residing at Attleboro, in the county of Bristol and State of Massachusetts, have invented a new, original, and ornamental Design for Pocket Lighters, of which the following is a specification, reference being had to the accompanying drawing, forming part thereof.

Referring to the drawing,

Fig. 1 is a perspective view of a pocket lighter showing my new design.

Fig. 2 is a top plan view of the same.

I claim,

The ornamental design for a pocket lighter as shown.

ALFRED F. REILLY.

June 28, 1927. A. F. REILLY Des. 72,965

POCKET LIGHTER

Filed March 9, 1927

Fig. 1.

Fig. 2.

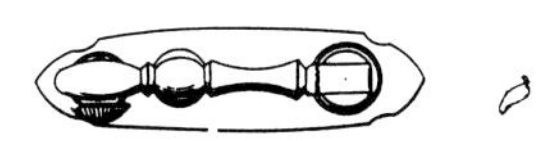

Inventor

A.F. Reilly,

By Robt. E. Barry

Attorney

Oct. 29, 1935. C. J. LAGERHOLM 2,019,432

LIGHTER CONSTRUCTION

Filed Aug. 4, 1928

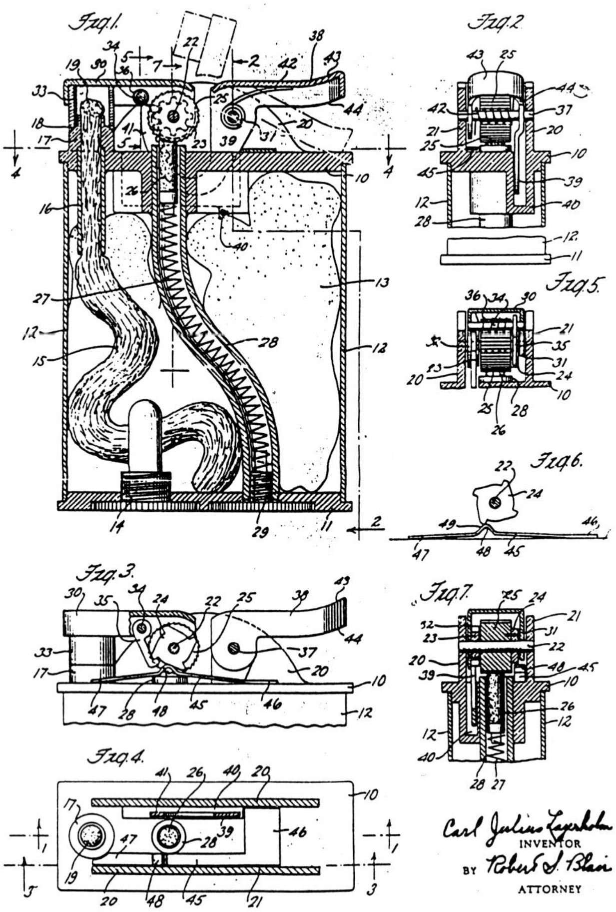

Carl Julius Lagerholm
INVENTOR
BY Robert S. Blair
ATTORNEY

Oct. 29, 1935. C. J. LAGERHOLM. 2,019,433

PYROPHORIC LIGHTER

Filed Oct. 25, 1930 2 Sheets-Sheet 1

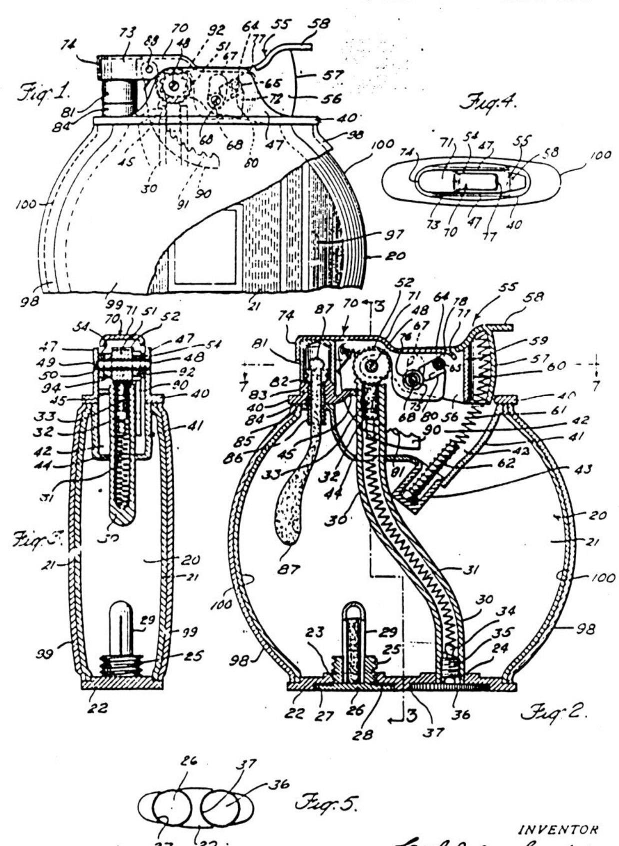

INVENTOR
Carl Julius Lagerholm
BY Darby, Blair & Curtis
ATTORNEYS

Oct. 29, 1935. C. J. LAGERHOLM 2,019,433

PYROPHORIC LIGHTER

Filed Oct. 25, 1930 2 Sheets-Sheet 2

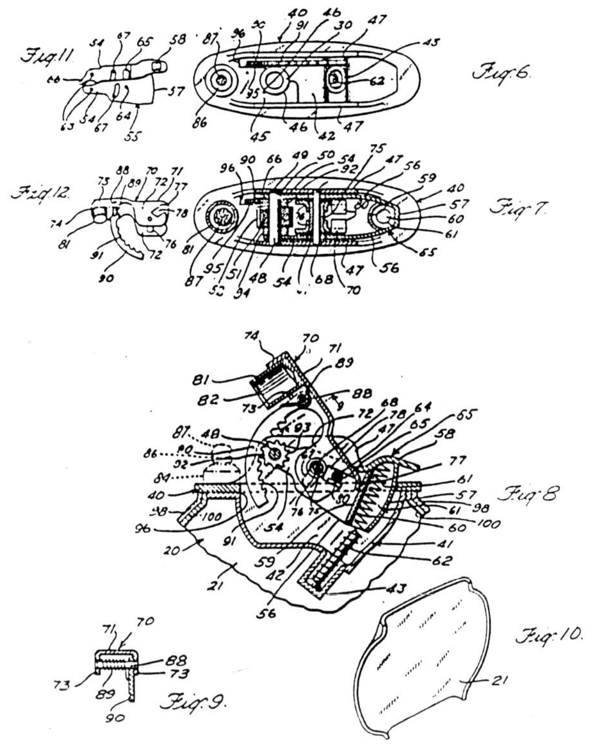

INVENTOR
Carl Julius Lagerholm
BY
Jenney, Blair & Curtis
ATTORNEYS

Oct. 29, 1935. A. F. REILLY 2,019,434

LIGHTER CONSTRUCTION

Filed Feb. 20, 1932

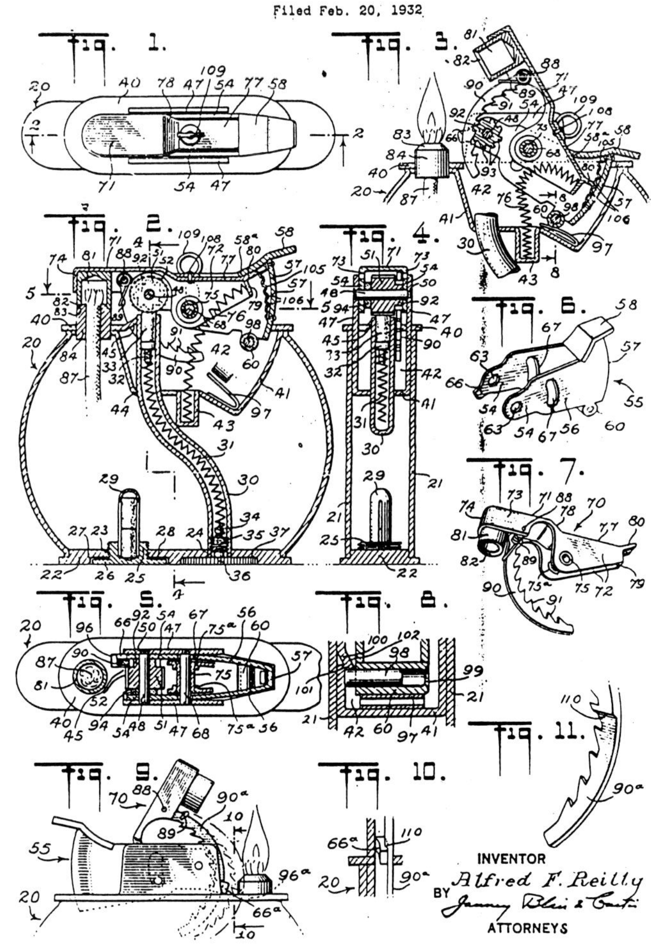

INVENTOR
Alfred F. Reilly
BY
Jenney, Blair & Curtis
ATTORNEYS

Oct. 29, 1935. R. S. BLAIR 2,019,435

PYROPHORIC LIGHTER

Filed Dec. 12, 1932 3 Sheets-Sheet 1

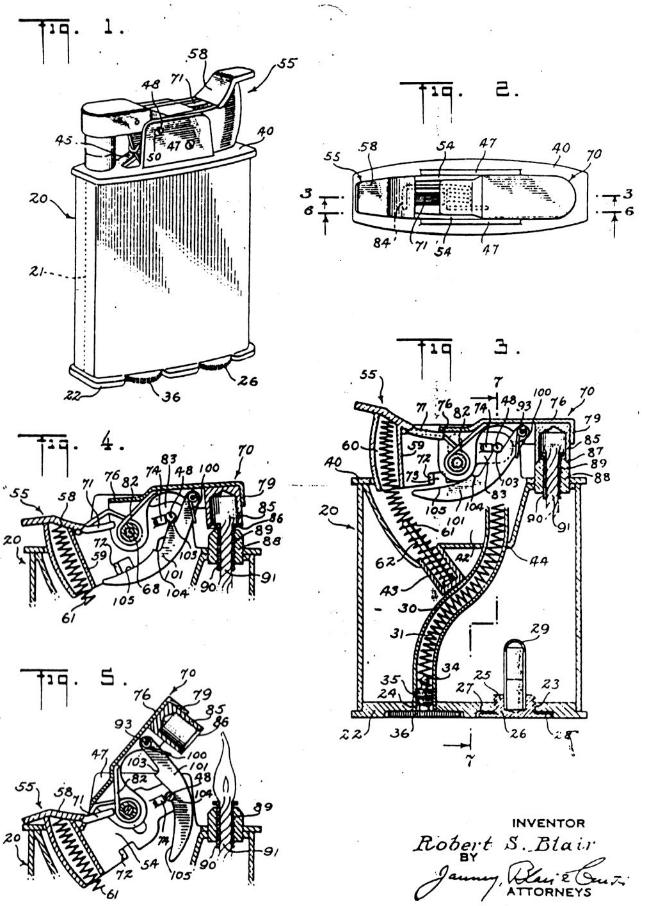

INVENTOR

Robert S. Blair

BY

Janney, Blair & Curtis

ATTORNEYS

Oct. 29, 1935. R. S. BLAIR 2,019,435

PYROPHORIC LIGHTER

Filed Dec. 12, 1932 3 Sheets-Sheet 2

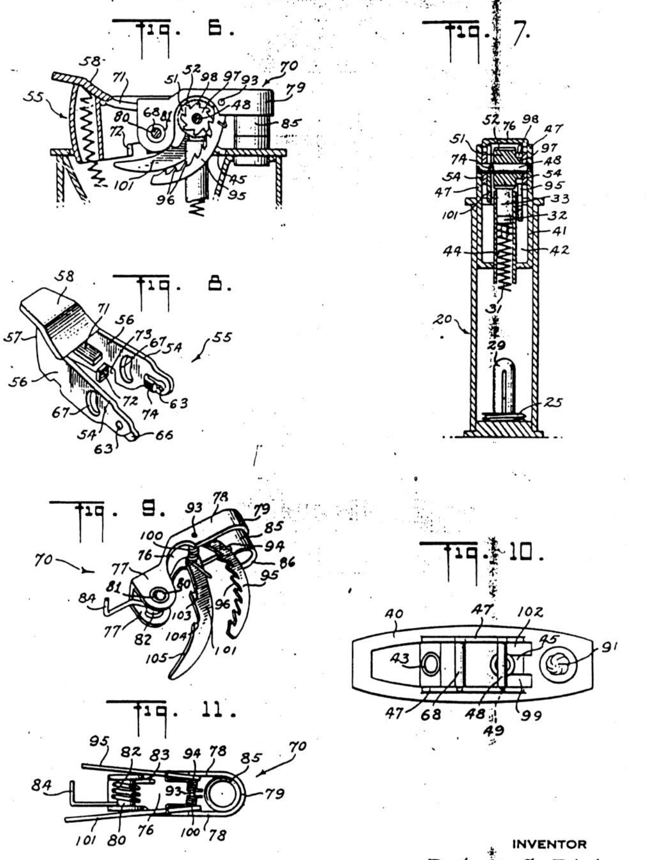

INVENTOR

Robert S. Blair

BY

Janney, Blair & Curtis

ATTORNEYS

Oct. 29, 1935. R. S. BLAIR 2,019,435

PYROPHORIC LIGHTER

Filed Dec. 12, 1932 3 Sheets-Sheet 3

Fig. 12.

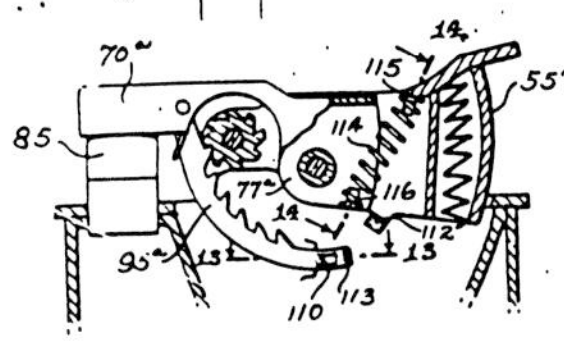

Fig. 15.

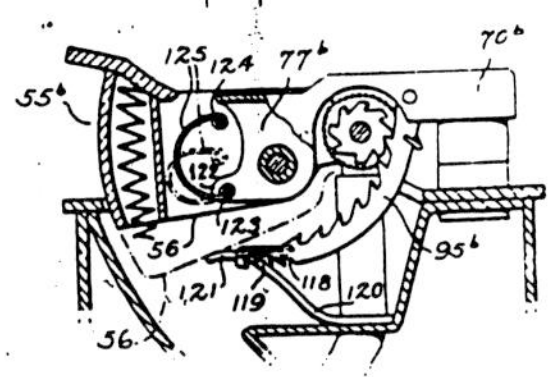

Fig. 13.

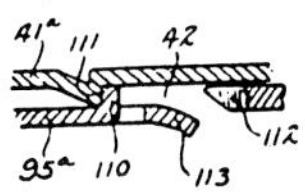

Fig. 14.

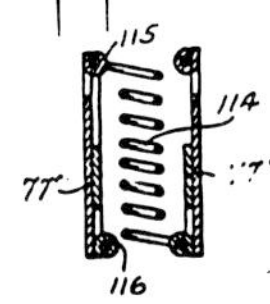

Fig. 16

Fig. 17.

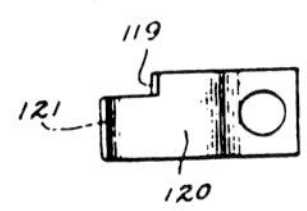

Fig. 18.

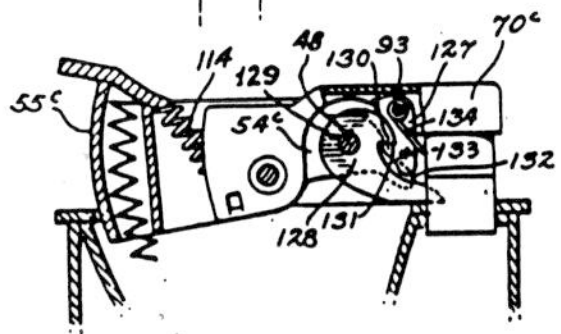

Fig. 19.

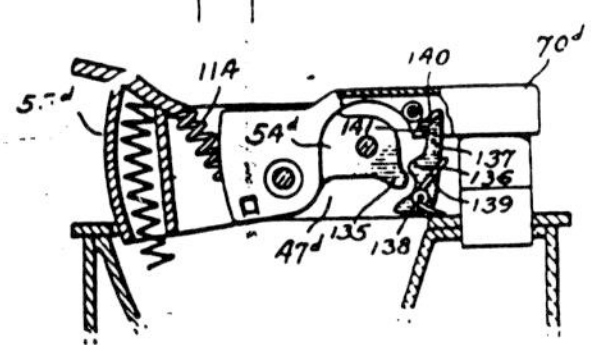

Fig. 20.

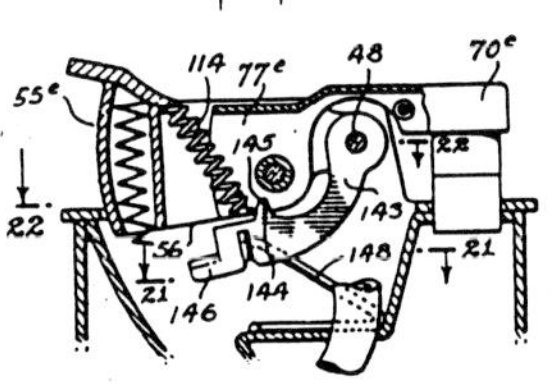

Fig. 21.

Fig. 22.

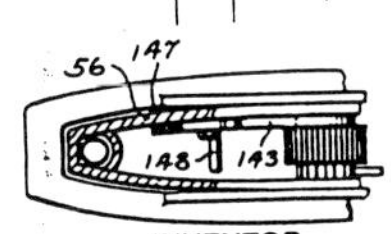

INVENTOR

Robert S. Blair

BY

ATTORNEYS

Oct. 29, 1935. T. M. GIBSON 2,019,436

LIGHTER MECHANISM

Filed Aug. 16, 1933

Fig. 1.

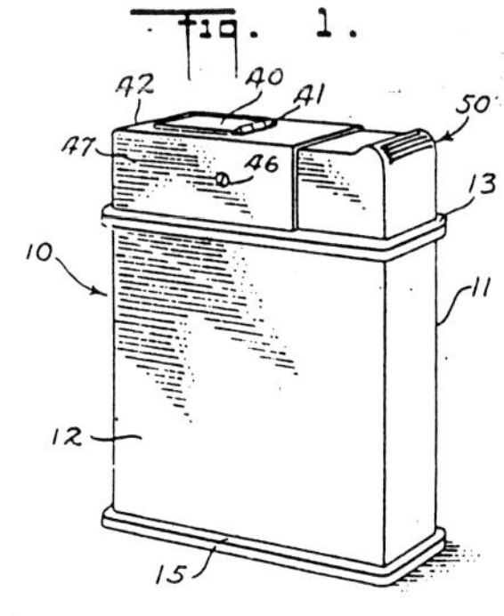

Fig. 2.

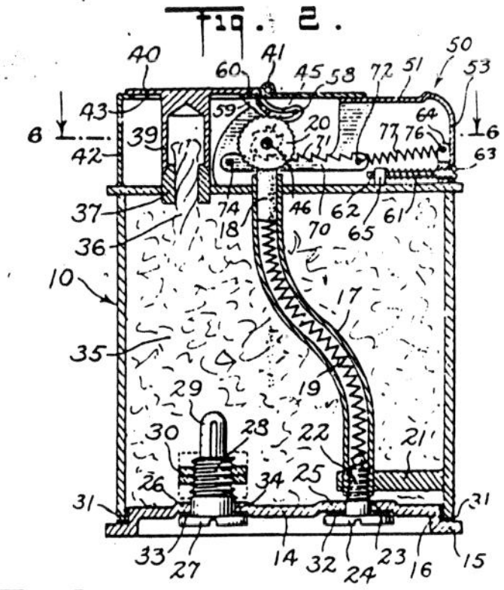

Fig. 3.

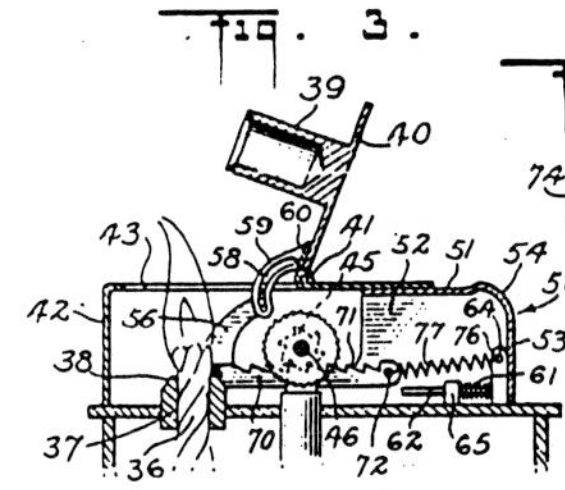

Fig. 8.

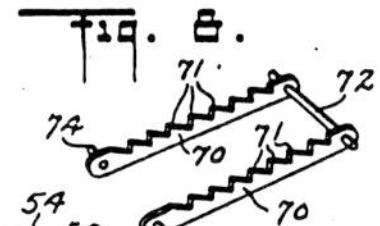

Fig. 7.

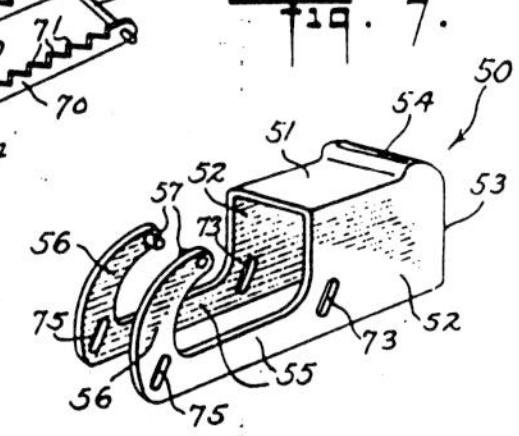

Fig. 4.

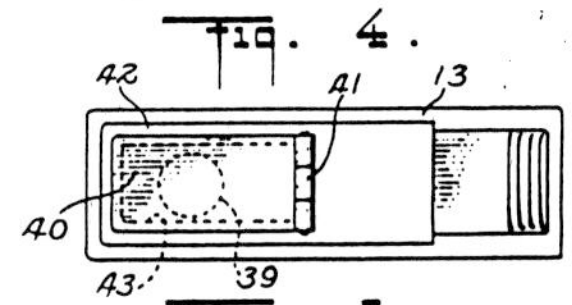

Fig. 5.

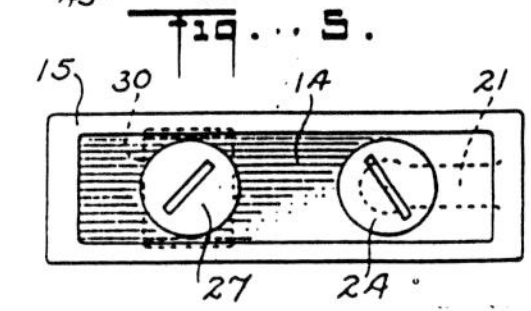

Fig. 6.

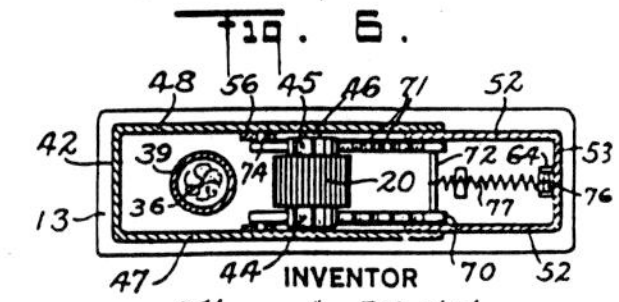

INVENTOR

Thomas M. Gibson

BY

ATTORNEYS

May 10, 1932. C. J. LAGERHOLM 1,857,200

CIGAR LIGHTER

Filed Dec. 13, 1928

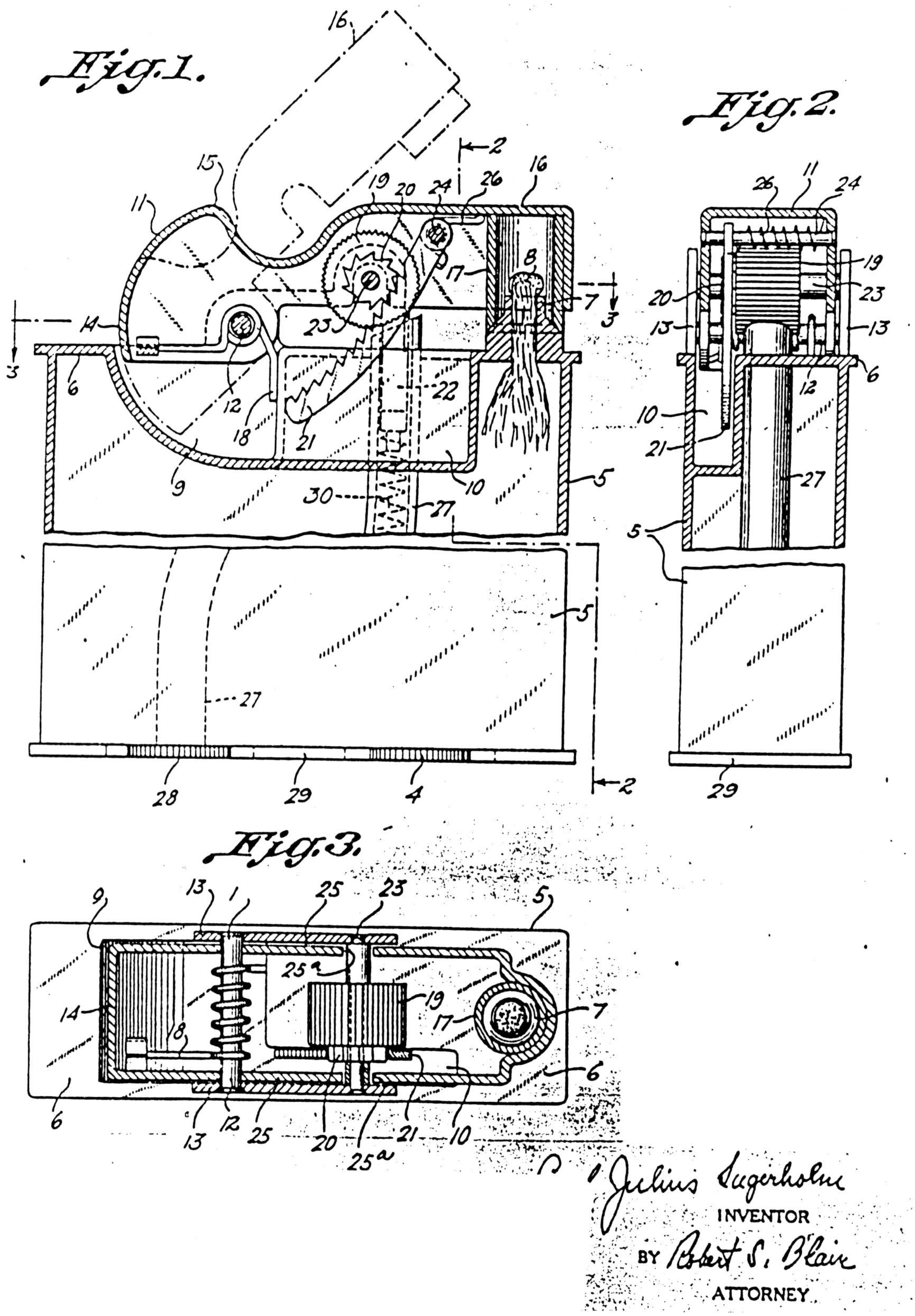

Resources

Imai, Kesaharu. *Lighters.* Tokyo: World Photo Press, 1997.

History of Massachusetts Industries, Vol. III, 1932.